Blacksmithing
TECHNIQUES

Other Schiffer Books on Related Subjects:

The Contemporary Blacksmith,
Dona Z. Meilach, ISBN 978-0-7643-1106-2

Ironwork Today 4: Inside and Out,
Catherine Mallette, ISBN 978-0-7643-4673-6

Decorative Architectural Ironwork:
Featuring Wrought & Cast Designs,
Diana Stuart, ISBN 978-0-7643-2192-4

Copyright © 2015 by Schiffer Publishing Ltd.

Library of Congress Control Number: 2015943720

Originally published as *Forja* © Copyright 2011
ParramónPaidotribo—World Rights. Published
by Parramón Paidotribo, S.L., Badalona, Spain.
Translated by Jonee Tiedemann.

Editorial management: María Fernanda Canal
Editor: Tomàs Ubach
Asst. editor and image archivist: Carmen Ramos
Text: José Antonio Ares
Projects: José Antonio Ares, Ernest Altés
Series design: Josep Guasch
Photography: Nos & Soto, Ernest Altés, Ares, Museu Cau Ferrat
infographics: Farrés il·lustració Editorial
Production director: Rafael Marfil
Production: Manel Sánchez

Typeset in Helvetica Neue

ISBN: 978-0-7643-4935-5

Printed in China

Published by Schiffer Publishing, Ltd.
4880 Lower Valley Road
Atglen, PA 19310
Phone: (610) 593-1777; Fax: (610) 593-2002
E-mail: Info@schifferbooks.com

For our complete selection of fine books on this and related
subjects, please visit our website at www.schifferbooks.com.
You may also write for a free catalog.

This book may be purchased from the publisher.
Please try your bookstore first.
We are always looking for people to write books on new and
related subjects. If you have an idea for a book, please contact
us at proposals@schifferbooks.com.

Schiffer Publishing's titles are available at special discounts
for bulk purchases for sales promotions or premiums. Special
editions, including personalized covers, corporate imprints, and
excerpts can be created in large quantities for special needs. For
more information, contact the publisher.

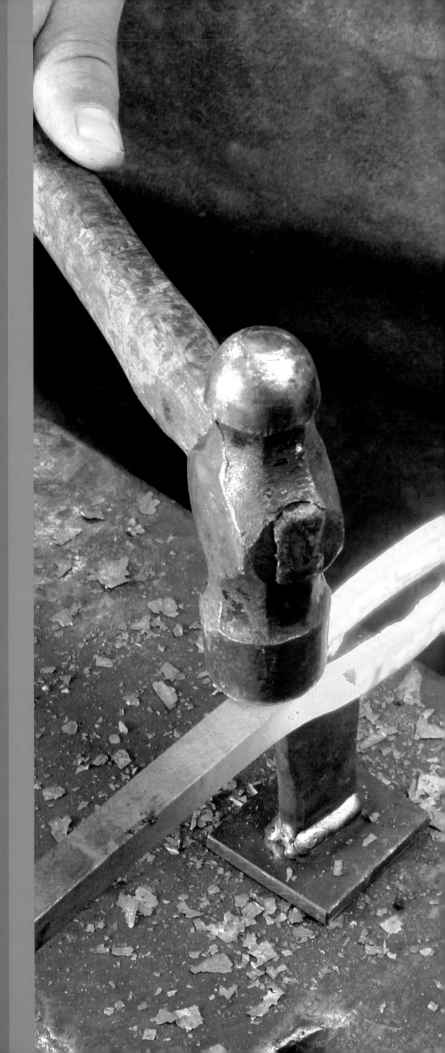

mithing
QUES

ARES

THE BA... STEP, COMPLETE WITH 10 PROJECTS

Schiffer Publishing

Schiffer
Publishing Ltd

4880 Lower Valley Road • Atglen, PA 19310

Materials
and tools

8 MATERIALS AND TOOLS

10 Raw materials
Iron and steel

12 Introducing the metals

14 Fuels
Coal

15 Gases

16 Protective materials
Patinas

17 Preparation of varnish
to protect metals

18 Basic tools
The forge

19 Firing up the forge

20 Color and temperature

21 Tongs

22 Anvil

24 Hammers, clamps, and vises

25 Other tools

26 Additional Tools
Tools and machines

30 Safety

Techniques
basic

Step by step

32 BASIC TECHNIQUES

34 General processes
Techniques for section changes

38 Techniques for
bending and curving

44 Techniques for
cutting and perforating

48 Special
cutting techniques

50 Heat treatments

52 Welding techniques
Arc welding

56 Oxyacetylene welding

58 STEP BY STEP

60 Cold-forged trivet

66 Door handle

72 Fireplace tongs

80 Grille with scrollwork

90 Wall candelabrum

102 Door knocker

110 Coat rack

116 Tricycle stool

122 Serpentine weathervane

130 Diàspora: a monumental forging

136 Gallery

142 Glossary

144 Acknowledgments

introd

This book focuses on the basic processes and methods of forging, and introduces the reader to the traditional techniques from a modern perspective. It covers classic skills like punching and bending, as well as modern processes like plasma cutting, heating with a torch, and using power tools. The goal of this book is to combine traditional methods with contemporary technological advances to help you take advantage of all the creative resources available.

The projects included in the "Basic techniques" section and in the "Step by step" section were selected so that one person working alone can complete them. This makes it easier for you to enter the world of artistic forging.

uction

Many blacksmiths make their own tools based on their projects' requirements. The "Basic Tools" chapter covers the essential tools used for basic tasks, rather than listing all the many tools that can be purchased or made. For the tasks and basic projects in this book, only carbon steel will be used, so we will refer to it throughout simply as "steel." Other materials like copper, aluminum, or stainless steel aren't discussed here since working with them requires more experience.

Finally, a "Gallery" of works by both well-known and anonymous blacksmiths shows the creative possibilities of forging. We encourage you to go ahead and experiment—use the different techniques presented here without worrying about making mistakes, and enjoy building your skills.

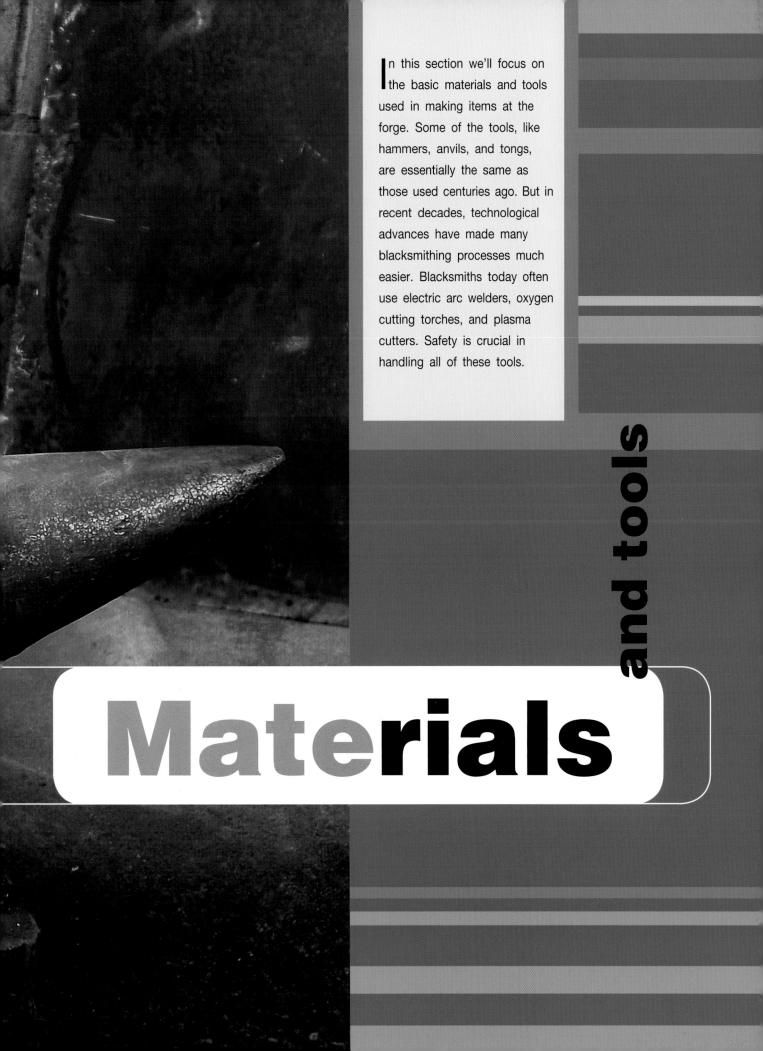

In this section we'll focus on the basic materials and tools used in making items at the forge. Some of the tools, like hammers, anvils, and tongs, are essentially the same as those used centuries ago. But in recent decades, technological advances have made many blacksmithing processes much easier. Blacksmiths today often use electric arc welders, oxygen cutting torches, and plasma cutters. Safety is crucial in handling all of these tools.

Materials and tools

RAW MATERIALS

Iron and steel

CHARACTERISTICS

The main raw material of the blacksmith's workshop is steel, specifically low-carbon steel. It's the easiest raw material to shape with hammering and forging techniques.

Steel is sometimes called iron, but this is a misnomer. The difference between them is their carbon content. Iron is an element, and is very soft. Steel is an alloy of iron and carbon. The amount of carbon dictates the steel's hardness and its elasticity, allowing it to undergo heat treatments such as tempering. If iron contains more than 1.7% carbon, it becomes fragile and brittle. Because it isn't ductile and malleable, it must be melted. This material is often used for casting objects in a mold.

Steel worked in a forge must be easily shaped, and able to be stretched by hammering to form rods, including very fine ones, without breaking or cracking. It should be neither too hard nor too soft; its carbon content should be around 0.15%.

Typical cast door opener in the shape of a hand holding a ball. Casting is commonly used to make pieces in special molds.

Steel can be shaped after heating it in the forge and striking it with a hammer.

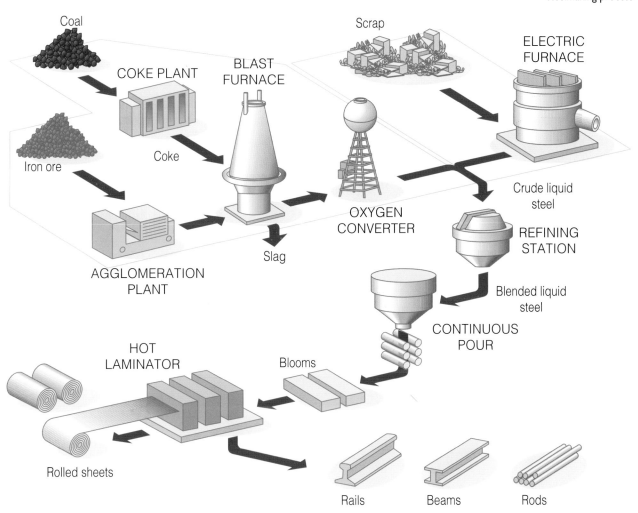

Coal

Scrap

COKE PLANT

BLAST
FURNACE

ELECTRIC
FURNACE

Coke

Iron ore

OXYGEN
CONVERTER

Crude liquid
steel

AGGLOMERATION
PLANT

Slag

REFINING
STATION

Blended liquid
steel

CONTINUOUS
POUR

HOT
LAMINATOR

Blooms

Rolled sheets

Rails   Beams   Rods

MANUFACTURING STEEL

Steel is produced from iron ore and scrap. Blast furnaces refine the pig iron (iron ore) into quality steel by forcing high-pressure oxygen over the molten metal.

The oxygen binds with the carbon and the unwanted elements and starts a process of high-temperature oxidation, which burns away the pig iron's impurities.

At the same time, fluxes such as lime are added to create a chemical reaction that produces heat (around 3000°F / 1650°C). Once the correct composition for steel has been achieved, it is poured into the continuous pour vat. This process can produce up to 300 tons of steel in about an hour. Electric furnaces are used to manufacture specialty steels and stainless steels because they don't use fuels that cause impurities.

The scrap must first be analyzed because its alloy content will affect the composition of the refined metal. Electric furnaces make it possible to control the temperature with great precision. A voltaic arc is generated between two large electrodes inside a hermetically sealed chamber. This produces heat of 3500°F / 1930°C for melting the metal. The precise quantities of the needed alloy materials are then added.

Materials and tools

Tubes manufactured cold from sheets of different thicknesses

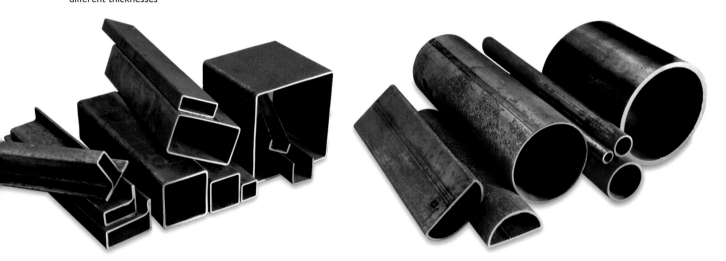

Introducing the metals

There are a great number of finished products on the market. Sheets and bars come in a variety of thicknesses and cross sections. These products are standardized based on their shape, finish, and intended use, and they are produced through cold forming or heat lamination.

COLD FORMING

Cold-formed steel shapes are made from a thin sheet ($1/32$ to $1/4$ inch / 1 to 6 mm). They are made in shaping machines that bend and fold the metal at ambient temperature. The sheet is bent and folded in various rollers until the shape of the desired profile has

been achieved. Then the profiles are joined through arc welding. During this process no lamination occurs since the thickness of the sheet is not changed. These steel shapes are commonly used in constructing furniture, metal handrails, and door and window frames.

Outline of the steps involved in manufacturing a round tube in an automatic shaper

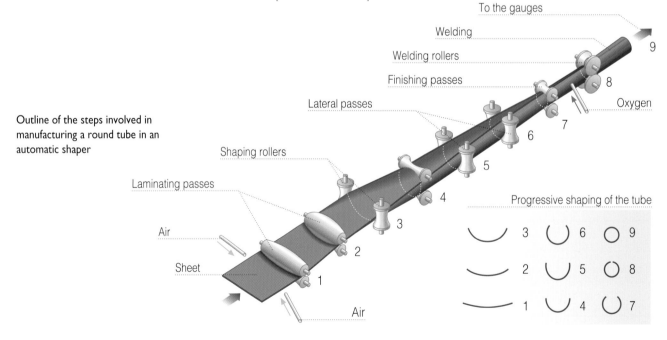

Various shapes produced by hot laminating

HOT LAMINATING

In hot laminating, the temperature of the metal is raised to 2192°F / 1200°C, and the metal is immediately run between two rollers, located one on top of the other and turning in opposite directions. While passing through them the metal is under tremendous pressure, which changes its structure.

This creates a continuous forging effect that improves the metal's characteristics. It eliminates any trace of weld or impurity that may have been produced during the melting, and the metal's ductility and resilience are improved. It becomes more resistant to breaking due to tearing, compression, and even torsion.

Due to their mechanical properties (elasticity), these shapes are most often used for structures such as bridges, buildings, towers for high-tension cables, and in the maritime industry.

The laminator's basic function.

Original grain

Elongated grain

New grain forming

Structure formed by new grain

Many functional works are made by forging and combining different shapes that are available commercially. *Hercules Fountain* (47 x 20 x 20 in / 120 x 50 x 50 cm), by Antoni Gaudí. Gardens of the Pedralbes Palace, Barcelona, Spain.

Materials and tools

Charcoal

FUELS

Fuels are needed to achieve the energy required to heat the metals until they reach the temperature for forging. They can be solid, like coal; liquid, like diesel; and gaseous, like oxygen, propane, and acetylene. In this book we will deal only with the fuels most commonly used in a blacksmith's shop: coal, acetylene, and propane.

Coal

The most commonly used coal is soft coal, crushed to pieces about the size of an almond. Moistening it with water allows a compact cave or vault to be formed in the forge where high temperatures can be created and maintained. This reduces the time required to heat the metal.

Sometimes charcoal is used as a fuel. It consists of carbonized hardwoods,

usually oak or holm oak. Charcoal is most often used for special tasks where precise temperature control is required, for example, making light alloys at a low fusion temperature.

There are other types of coal, such as coke, anthracite, and lignite, which are not as commonly used for forging. Anthracite is very expensive, and lignite has a limited heating capacity.

Soft coal

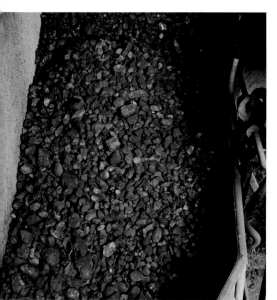

Coal arch formed by spraying water onto the hot coal

Comparison of flame temperatures produced by mixing oxygen with various gases

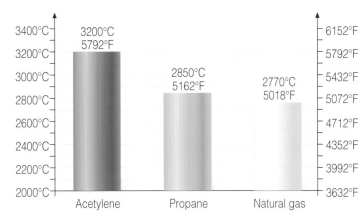

Gases

The most commonly used gases in the blacksmithing shop are oxygen, acetylene, and propane. They are sold in highly pressurized metal cylinders, which are labeled or distinguished by internationally standardized colors.

They are used to produce a flame that reaches a high temperature, by mixing welding torch oxygen and acetylene, or oxygen and propane. Oxygen can also be combined with domestic piped-in natural gas, but the temperature will be lower.

The flame is usually used to localize and limit the heating of the piece to be forged; this allows for bending, crimping, and twisting to be done without the need to work the piece inside the forge. It also avoids the inadequate heating of other sections of the piece, which might cause deformation.

Cylinders containing oxygen (black) and acetylene (red)

Localized heating with an oxygen and propane torch

Cylinders for storing oxygen (black) and propane (orange)

PROTECTIVE MATERIALS

Tools and products for producing a black patina: gas torch, linseed oil, and wax

Patinas

In order to protect metals from rust, a patina (varnish, wax, or paint) must be applied.

Patinas are applied to the surface with a cotton cloth or brush; the coating provides an aged or antiqued look. Before a patina is applied, the metal must be cleaned to remove all oils and any traces of rust.

The simplest patina consists of a mixture of distilled water and salt. With time this mixture creates a layer of rust that protects the metal. The metal then must be sealed by applying a varnish to form a thin, hard coating.

Black finish patina: This conserves the texture produced on a piece while working it on the anvil. Linseed oil is applied to the piece, and it is heated with a torch to produce the blackness desired. After the piece is cooled, it is rubbed with wax using a cotton rag. This seals the pores.

Graphite patina: This produces a gray protective patina. The graphite is dissolved in linseed oil to create a smooth paste, then a universal solvent is added to it. To darken or color the patina, various pigments can be added.

Waxes: Like varnishes, they polish and finish the surface of a piece. Depending on the desired shine, more or less pressure is applied to the cotton rag used for polishing.

Gum lacquer varnish: To prepare this varnish, mix 5¼ ounces (150 g) of gum lacquer scales with one quart (1 l) of denatured alcohol.

Paints: Paints are mixtures of pigments, coloring agents, and synthetic binders. They adhere well to metal, and can have an opaque, glossy, or matte finish.

Gum lacquer scales and denatured alcohol for making gum lacquer varnish.

Graphite patina made from linseed oil, graphite, black pigment, and universal solvent

Preparation of varnish
to protect metals

A layer of rosin and wax is applied with a brush to protect the metal from rust. This also evens out the colors, and provides a satiny base coat for easier application of the final finish.

1

2

3

1 Varnish is a mixture of turpentine, beeswax, and rosin.

2 Beeswax is sold in sheets and slabs. It is chopped into pieces so it will dissolve more easily.

3 Crystallized chunks of rosin are prepared for dissolving by wrapping them in a cloth and pulverizing them with a hammer.

4

4 Measure by weight one part beeswax and one part rosin. Place them in an enamel or glass container. Add three parts turpentine, and dissolve the mixture over hot water in a double boiler, producing a liquid.

5 The mixture thickens into a semi-liquid substance as it cools.

5

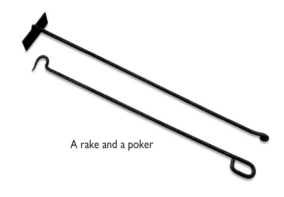
A rake and a poker

BASIC TOOLS

The **forge**

Old forge with bellows

PARTS OF THE FORGE

The forge consists of three parts: a base, a fan, and a hood. The base, or hearth, is where the coal is burned and where the metal pieces are heated. It is placed at a certain height above the floor, and covered with refractory brick. The fan is usually below the base and is run by an electric motor; it provides a steady stream of air to the fire in order to stoke it up. The exhaust hood vents the smoke and gases away through a chimney.

The coal bin may be located next to the forge, although more often it's placed outdoors to make more effective use of the interior workspace. A complete set of forging equipment includes a poker for arranging the hot coals and a container of water to sprinkle onto the coal or to cool the heated pieces.

A fully-equipped forge setup

Fan and ducts at the base of the forge

Portable forge

Firing up the forge

This is one of many methods used to light the forge. It is not the only way, and different people use different methods.

Once the fire has been lit, wait until the sulfurous gases that the coal gives off are consumed before heating the steel, as the gases might harm it.

1 Before lighting the forge, check and clean the fan's air supply duct to make sure there are no leftover pieces of coal or ashes inside it. Place coal in the hearth of the forge. Using a rake, open up a space in the center of the base just above the air outlet to facilitate air delivery from the fan and ignition of the fire.

2 Kindle the fire in the middle of the opening using paper and wood shavings, and light with a match, cigarette lighter, or another convenient lighter.

3 Gradually, use the fan to supply air to the fire through the vent. At the same time, add coal to the fire.

4 A column of dense smoke indicates that the coal has caught. Because the smoke contains sulfurous compounds that can damage the metal, don't put the steel into the hot coals until this smoke has burned off.

5 The fire is at the right point for heating pieces. To avoid rust formation on the steel, do not place it directly above the air vent.

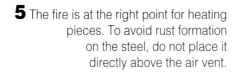

Materials and tools

Color and temperature

GENERAL INFORMATION

Metal is heated while keeping in mind its mass and its physical and chemical properties.

If the metal is heated without following the right procedures, there is a risk that the metal's essential characteristics, for example its rust resistance, may be changed. The speed of the heating process depends on:

- the physical and chemical properties of the metal. A more conductive metal will take on heat more quickly.

- the metal's mass and the relationship among its dimensions. A piece with a greater surface area will absorb more calories and heat up more quickly, even if it weighs the same as or is made of the same material as a different piece.

- the temperature of the coals in the forge.

A dimly lit room allows for better viewing of the color stages the metal passes through as it is heated in the forge.

COLOR AND TEMPERATURE

Heating metal produces color changes that vary according to the temperature, and thereby help you to visually determine the temperature. For example, dark cherry red indicates that the temperature is up to about 1562°F (850°C), and orange corresponds to about 1832°F (1000°C).

Judging the color depends on the composition of the steel and the light in which it is examined. In order to most clearly see the colors of the heated steel, it is helpful to examine it in a dimly lit workshop.

The heating of the piece must be done gradually, turning it in the hot coals now and then to allow for an even heating of all of its surfaces and to achieve heating of the core, especially important for thick cross sections. Also, the piece should be placed in the coals far enough away from the direct air intake of the vent to avoid surface oxidation, but also deep enough in the coals so that it doesn't get too hot in the reduction area.

Relation between color and temperature in heated steel

Temperature	Color
2552°F (1400°C)	pale white; upper limit for forged steel
2372°F (1300°C)	white yellow
2192°F (1200°C)	pale yellow
2012°F (1100°C)	yellow
1832°F (1000°C)	orange
1742°F (950°C)	yellowish red
1652°F (900°C)	light red
1562°F (850°C)	red
1490°F (810°C)	light cherry red
1472°F (800°C)	
1400°F (760°C)	cherry red
1364°F (740°C)	dark cherry red
1292°F (700°C)	Minimum forging temp.
1256°F (680°C)	dark red
1148°F (620°C)	brownish red
1112°F (600°C)	
1022°F (550°C)	dark brown
932°F (500°C)	
752°F (400°C)	
680°F (360°C)	gray
644°F (340°C)	bluish gray
608°F (320°C)	light blue
572°F (300°C)	blue
554°F (290°C)	dark blue
536°F (280°C)	purple
518°F (270°C)	reddish purple
500°F (260°C)	bronze
482°F (250°C)	brown
464°F (240°C)	dark straw
446°F (230°C)	light yellow
428°F (220°C)	straw
392°F (200°C)	light yellow

Examples of tongs

Tongs

Tongs are a basic tool for blacksmithing. They are made up of two articulated bars and a hinge that separates them into the handles, by which they are held, and the jaws. The jaws, used to secure the pieces as they are worked, come in many shapes and configurations, allowing them to be useful on different shapes. They can be flat, rounded, grooved, L-shaped, and so forth. Tongs are used to insert, manipulate, turn, and remove pieces from the hot coals. They help to prevent working too close to the fire and inadvertently touching the hot steel, and are especially useful for heating small pieces.

Heat can warp the tongs, and hard use can open the jaws to an ineffective point. They require regular inspection and repair to ensure a good, safe grip on the steel being worked.

Tong jaws holding and manipulating metal

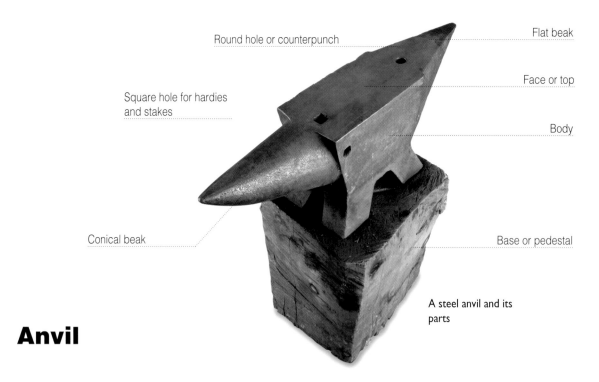

Round hole or counterpunch

Square hole for hardies and stakes

Conical beak

Flat beak

Face or top

Body

Base or pedestal

A steel anvil and its parts

Anvil

The proverbial tool of the blacksmith's workshop is made from a block of solid forged steel, upon which metal is forged with blows of the hammer. The anvil generally weighs from 75 to 440 pounds (34 to 200 kg), depending on the type of work for which it is intended.

The top surface of the anvil usually has two holes: a square one to accommodate forging accessories,

and a round one called the counterpunch or pritchel hole. This is used for punching holes and making bends. The anvil typically has one or two ending points called beaks, one conical and the other flat or pyramid-shaped. These are used for stretching, bending, and folding.

Anvils are placed on sturdy, very stable bases, which absorb the blows and reduce the rebound effect. The

height of the anvil should allow the blacksmith to keep his back almost straight while working, and should reach the knuckles of the standing blacksmith's hands.

The anvil should be close to the forge so that the blacksmith can stand between the two. This allows him to take the heated steel from the hearth, and turn immediately to the anvil to work it.

Forging on the flat beak of the anvil

Forging on the rounded, conical beak of the anvil

Stakes (A), hardies (B), and
forging cones (C)

ANVIL ACCESSORIES

Sometimes additional tools for the anvil are needed for delicate or specialized tasks. They are used for forging pieces or parts that are either cold or hot. These accessories have a square peg on the bottom that fits into the square hole of the anvil to keep the accessory from turning or moving while in use.

Many of these accessories are made by the blacksmith himself to meet the requirements of the job at hand. Some of them are so specialized that they are only used once, then are recycled or discarded.

Stakes: make it possible to hammer the steel and bend it, flatten it, or groove it.
Hardies: used for cutting steel pieces, either hot or cold. They fit into the square hardy hole with a cutting edge facing upward. Metal is held on top of the cutting edge and then struck with a hammer.

Bending forks: are used for bending and curving steel on the anvil.

Forging cones: used for bending hoops or rings and for curved or square pieces.

Bending fork

Bending a heated strap
in a bending fork

A variety of forging hammers

Locking pliers and
a manual screw clamp

Hammers, Clamps, and Vises

Ball and peen hammers: essential tools for forging work with hot or cold metal. The most common types are ball hammers, with a nearly spherical ball on one end, and peens, which have a rounded face. They weigh between roughly 2 and 4½ pounds (9 and 2 kg), depending on the type of work being done.

Blacksmith's vise: the most common holding device in forging work. This is a robust tool made from forged and tempered steel to resist great pressure. It consists of a movable jaw held away from a fixed jaw. Blacksmith's vises are normally attached to a workbench with bolts and clamps. Sometimes special devices are used when installing them so that there is open space around the bolts, which facilitates working with large pieces.

Bench vise: made from cast or forged steel. The vise is used to hold pieces firmly in order to file, cut, bend, or perforate them, among other operations.

Hand vise: makes it possible to hold pieces of varying thicknesses. Their portability makes them essential in the blacksmith's shop.

Locking pliers: very handy for manipulating and working with small items that are difficult to hold. Locking pliers exert pressure by adjusting the opening between the jaws to the thickness of the piece.

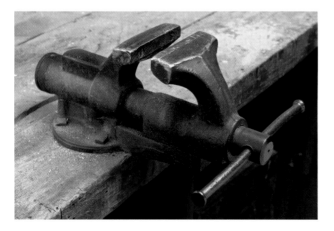

Bench vise

Blacksmith vises

Scrolling wrenches

Forks

Punches

Other **tools**

Scrolling wrenches: used as levers on steel forms to bend, fold, or twist them. The blacksmith makes them based on the shape he needs to create.

Bending forks: U-shaped tools made by the blacksmith to impart a curve of a certain radius. They can be made from round bar stock bent into a U shape or simply two pieces of the same bar welded to a base.

Punches: made from round or square rods of hardened steel with a sharpened point on one end, punches

are used to make holes in red-hot metal by striking the other end with a hammer.

Templates for curves: used for making the same shape several times. The blacksmith makes them from sheet stock with the desired shape welded onto a steel base.

Nail heading tool: used for making forged nails and rivets. It features round or square holes of various sizes. It is used over the round hole in the top of the anvil.

Rivet set and base plate: used to shape the heads of rivets. The rivet set is round in cross section and has a hardened and tempered opening in the rear. The mouth has a semicircular opening where the heads of the rivets are placed for swaging (tapering) to shape. The base plate fits into the square hole of the anvil. The head of the forged rivet is placed on it so that it will not be distorted while the other end is being formed.

Template for curves

Nail headers

Rivet sets and base

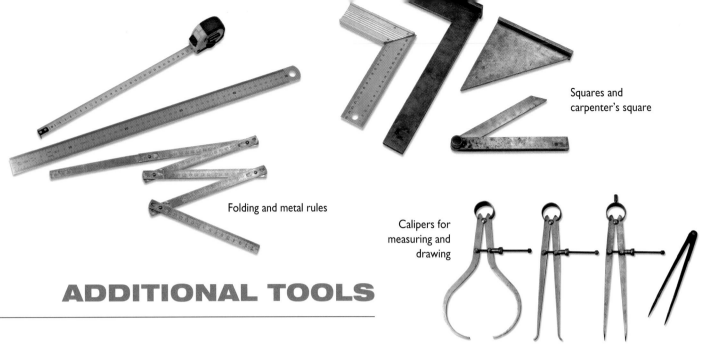

Squares and carpenter's square

Folding and metal rules

Calipers for measuring and drawing

ADDITIONAL TOOLS

Tools and **machines**

On the following pages we'll describe the additional tools and machines used for cutting, perforating, and welding steel. Measuring, tracing, and marking tools round out the workshop inventory needed for the most common forging tasks.

Metal and folding rules: made of stainless steel, they are very helpful in drawing lines on metals. A folding rule is a strip of thin, flexible steel used to measure flat surfaces and large objects. Rules commonly come in lengths of 3 to 16½ feet (1 to 5 m) or more, in contrast to measuring tapes, which are used for measuring lengths over 30 feet (10 m).

Squares and adjustable squares: used for marking and checking right angles. Never check a large surface with a small square because possible errors aren't visible beyond the end of the tool. The adjustable square is useful for transferring and marking certain angles, and for checking and confirming them. It consists of two jointed pieces that facilitate drawing any angle.

Calipers and compasses: made of steel and fitted with hardened ends, these are either simple or spring-operated. In addition to drawing arcs, compasses are useful for transferring measurements during forging work. Measuring compasses, which have

either concave or convex arms, are used for comparing and checking the interior or exterior of pieces.

Center punch: a steel tool with a hardened and tempered point, it is used for marking guide points for drawing arcs and circumferences. It is also used to mark the center that will guide a drill bit.

Scribe or scratch awl: a steel rod with a very sharp, hardened point. It is used to mark straight lines with the help of metal rules and squares.

Center punches

Scribes

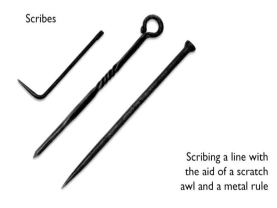

Scribing a line with the aid of a scratch awl and a metal rule

Cross section of files: triangular (A), flat (B), square (C), knife (D), round (E), and half-round (F)

Different types of files

Files: used to eliminate the rough edges produced by cutting, and to smooth out surfaces. Files are distinguished by their shape, length, and coarseness (fine or coarse).

Frame saws: tools composed of a steel frame or bow that holds and tightens a saw blade. Two clamps, one at each end of the bow, control the tension on the blade.

Chisels: wedge-shaped cutting instruments made from hard steel with a tempered edge. Also called cold chisels.

Burins: have a tapered cutting edge that runs at an angle to the body. They are used for making grooves and flutes.

Mechanical shear: this sturdy tool is used for cutting flat shapes of various thicknesses. Some are capable of cutting T- or L-shaped stock. The lever operates the shear and the gearing multiplies the force exerted on it. The longer the lever, the greater the force.

Burins and chisels

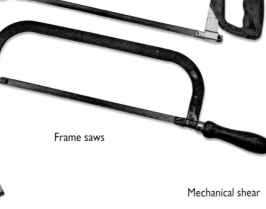

Frame saws

Mechanical shear

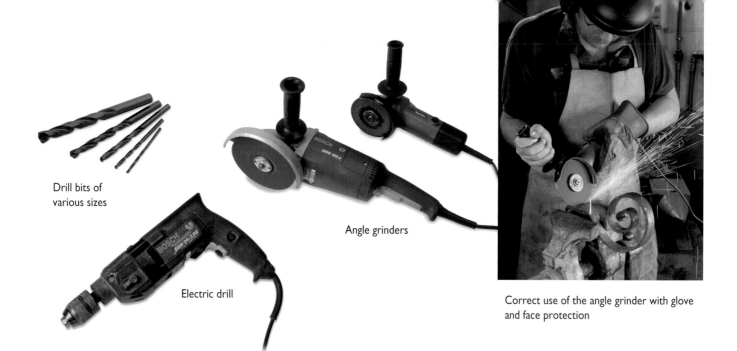

Drill bits of various sizes

Electric drill

Angle grinders

Correct use of the angle grinder with glove and face protection

Electric drill: consists of an electric motor that continuously turns a chuck, where the drill bit is secured.

Drill bits: cylindrical rods of hardened and tempered steel in varying diameters. They have spiral flutes with a cutting edge that start at their tip. The harder and more resistant the material to be drilled, the greater the angle required on the tip.

Angle grinder: portable electric tool that spins abrasive discs at 6,000 to 10,000 rpm (rotations per minute). The discs get worn down as they cut or abrade.

Jigsaw: portable electric tool used for quickly doing straight or curved open work. The motor imparts an elliptical movement to a saw blade attached to a movable axle.

Band saw: electric bench tool with a continuous saw blade that produces a continuous cut. It has adjustable speeds for handling different metals, and can make 90° to 45° angle cuts.

Electric bench grinder: a machine tool used to sharpen tools and remove rough edges. It has an electric motor and abrasive wheels at each end, usually one each of fine and coarse grits.

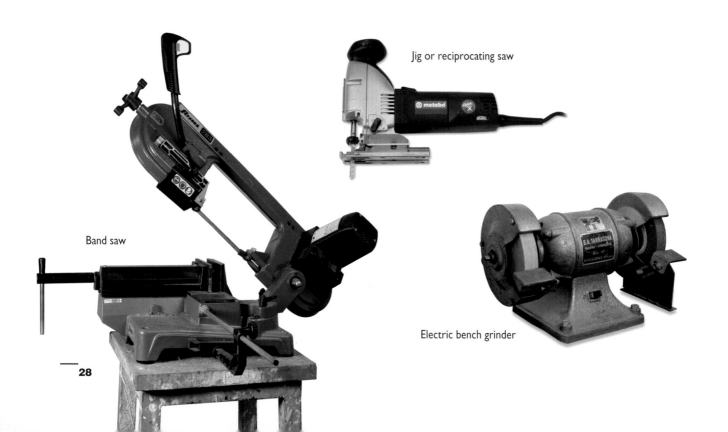

Jig or reciprocating saw

Band saw

Electric bench grinder

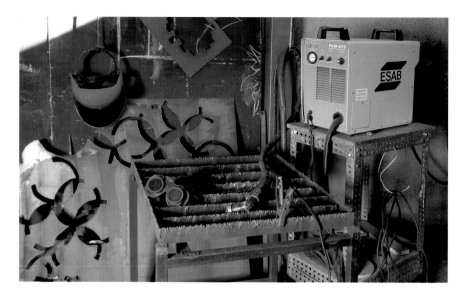

Oxyacetylene welding equipment

Plasma cutting
equipment

Plasma cutting equipment:
a high-frequency generator produces
an electric arc between the nozzle
of the torch and the ground clamp.
A gauge on the front measures the
outlet for the air necessary for plasma
formation. It also has a control for the
current output, which is adjustable
from 10 to 60 amps.

Oxygen cutting equipment:
consists of two tanks containing
pressurized gases (acetylene,
propane, and oxygen) and pressure
regulators indicating and maintaining
ideal working gas pressure. A cutting
torch mixes the gases and combusts
and removes the steel.

Oxyacetylene welding
equipment: also consists of two
tanks that contain pressurized gases.
However, the torch is different from that
of the oxygen cutting equipment. Its
nozzle is designed for joining steel by
melting it.

Arc welding equipment with
coated rod: this electric welder
(also called a "stick welder") uses a
transformer that modifies the electric
energy into a constant current.
A conductive electrode (with a
consumable flux-coated metal rod in its
center) and a ground clamp make up
the welding circuit.

Oxygen and propane
cutting equipment

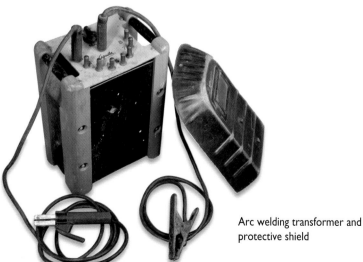

Arc welding transformer and
protective shield

Materials and tools

Metal hooks hung on a wall store bar stock until it is used.

SAFETY

SHOP ORGANIZATION

When a few organizational and safety measures are taken, you can enjoy a comfortable, efficient work area and avoid accidents.

The most frequently used tools, such as hammers, tongs, and portable power tools, should be kept in specifically designated places (on racks, on hooks, in cabinets, etc.) so that the tools are easy to access, and easy to store properly after use.

Cut-offs and scraps from forging work should be kept in containers that make it convenient to locate them and recycle them in future projects. Avoid the temptation to keep or accumulate so much material that your storage space is filled to overflowing. Only keep materials that are likely to be useful in the foreseeable future.

A metal container made by the black-smith is used to store leftover pieces for use in future projects.

Hammers with nicked or damaged striking surfaces should be discarded to prevent metal shards from flying off and causing serious injuries.

Hearing protection muffs
and foam earplugs

Basic personal protection
equipment for working
at the forge

PERSONAL SAFETY EQUIPMENT

Personal safety equipment is an essential feature in the workshop. Rotary grinders, electric saws, oxygen torches, and plasma cutters all produce shavings or red-hot particles.

A leather apron, gloves, and protective eye wear keep any particles from coming into contact with the body. To avoid burns, always wear gloves while welding, grinding, and cutting. Protective shoes or boots shield the feet from accidental blows.

Finally, wear ear protection or earplugs to dampen the noise produced by hammering on the anvil and by the operation of machinery in general.

In this section we explain the basic, traditional techniques of blacksmithing, along with other techniques that are relatively new such as cutting with oxygen and a stream of plasma. By combining traditional techniques with new technology, you can more efficiently create forged items without any loss of creativity. We've chosen techniques that can be used by a person working alone, which will make it easier to become familiar with them.

basic Techniques

GENERAL PROCESSES

Here are the most common basic techniques and methods, grouped according to the effect they have on the metal. They can be completed without the help of another person in the shop. All of these techniques require continuous practice to hone your skill.

A section on welding techniques is also included since they are indispensable in any blacksmithing shop, no matter how modest.

Techniques for **section changes**

Techniques that enlarge, reduce, or modify the original section of a commercially produced shape include upsetting, drawing out, tapering, laminating, and forging nails.

Upsetting

Upsetting is done to broaden and shorten the end of the steel for making such things as nail heads and ball-shaped ends on bars.

To upset a piece properly, the area to be broadened is heated until it is a light orange color. The heat must be even and must reach the center of the steel. At the same time, the part of the rod that is not to be upset must be cooled by sprinkling water on it; this prevents it from bending under the hammering. However, it is normal for the bar to bend a little just above the flared area. This must be controlled and corrected during the whole process.

Once the piece is heated, it is hammered in a direction parallel to the steel that is to be upset. The blows can be applied in different ways based on the thickness and length of the piece. The heated area of the bar may be hammered on the anvil, held in the blacksmith's vise, or struck on the face of the anvil. The upsetting is usually repeated several times.

Left: Upsetting the head of a rivet on the blacksmith's vise. Here the end of the shaft is hammered as it's heated by the torch.

Above: Upsetting the bent end of a rod by hammering on the face of the anvil. This way the area of the bend is thickened to counteract the thinning produced by bending the rod.

Another way to upset the steel is to strike the bar directly on the face of the anvil.

DRAWING OUT A TENON

This is the process of stretching a rod on just one side.

1 To begin drawing out the steel, place the appropriately heated rod on the edge of the anvil face and strike it. This forces the material to displace and marks the start of the stretch.

1

2

3

Drawing Out

The process of drawing out involves lengthening the material at the same time the thickness is reduced. To achieve this, the area to be drawn out is heated to a bright red and hammered, either on some part of the anvil such as the round horn or the face, or on a hardy. The drawing out is usually done in several phases. As the steel stretches and thins, be careful to avoid overheating it and ruining it in the hot coals.

2 Continue drawing out the steel on a flat hardy by striking the heated end with a hammer. Repeat this operation until the desired stretching has been achieved.

3 Heat the drawn-out end very carefully to apply the finishing touches.

Sequence of the drawing-out process using a tenon

Detail of a gate finial made by Antoni Gaudí at the Güell winery

Tapering

Tapering finishes off the section of the commercially supplied shape to a point or a thin line. This reduces the initial section while simultaneously producing a lengthening in the material. To taper, heat the steel to bright red and hammer it on the face of the anvil. Repeat the operation several times if necessary, taking special care in heating the tapered point.

Process of tapering a square rod

Tapers made by Antoni Gaudí on the entry gate of the Güell Winery

Process of laminating a round rod

Laminating

Laminating reduces the initial section by hammering out the material to increase its width. The piece must be heated in proportion to its thickness, keeping in mind that it will be easier to flatten if the temperature is higher. Thus, the hammering out is best done with the steel at light yellow rather than dark red.

Modern door knocker in the Gràcia neighborhood of Barcelona. The impression in the leaf was created through laminating.

Sequence in the process of laminating a round rod. Note that the operation of hammering out is done in several stages to ensure the right temperature for working the steel.

Forging nails

Various Basic techniques are combined to create forged nails, including upsetting, laminating, tapering, and trimming, among others.

Tapered, trimmed, upset, and shaped: the sequence for making a nail

A manual heading tool with a handle, and one or more holes in the shapes and sizes appropriate for the nails, is required. The process involves drawing out and tapering the rod to create the shaft of the nail, and trimming it on a hardy attached to the anvil. The head of the nail is created through upsetting in the heading tool. The piece is heated several times during the forging process.

MAKING A NAIL

The following sequence begins with the end of the nail previously hammered out and tapered, as demonstrated on the previous pages.

1 A heated rod, tapered in advance, is struck gently on a hardy in the anvil to mark the end where the nail will be cut or trimmed.

2 Then the rod is inserted into the hole in the heading tool held over the round hole on the face of the anvil. It is bent over until it breaks free where the cut was marked on the hardy.

3 The head of the nail is upset on the heading tool by shaping the steel at bright red heat.

4 The nail, held with appropriate tongs, is worked and shaped using the hammer.

Renaissance nail on the door of the Church of Santa Maria la Major in the town of Prades, Spain

Basic techniques

Techniques for **bending and curving**

Techniques such as rolling and making clips or spiral scrolls are all variations on the curving and bending processes. These are very expressive techniques commonly used in making forged items. Some can be worked cold without having to heat the steel, as long as the thickness of the material allows it. However, for the best results, it's usually necessary to use heat.

Cold bending and twisting

There is no heat involved in cold bending and twisting the steel. Usually, the thickness of the metal puts limits on the process.

Bending involves imparting a curve, and twisting creates a spiral in the steel. The blacksmith uses tools that multiply force such as bending forks, hand scrolling wrenches, and various tools for clamping both to the face of the anvil and the blacksmith's vise. Curves and twists can also be done with cold steel by hammering on the rounded anvil beak or right on the blacksmith's vise.

Using a U-shaped form on the anvil face to bend a piece of strap. The hammering bends the strap at the point located between the ends of the shape.

The process of bending the strap on the rounded horn of the anvil

Bending in a fork held between the jaws of the blacksmith's vise. In this case, a scrolling wrench is used for better control of the curve.

Using a scrolling wrench to cold bend a strap

Detail of a grille entirely made through cold bending, by Ares

Hot bending and twisting

Heating the steel facilitates and simplifies the forging process and achieves the greatest plasticity. The metal must be heated to bright red through to the center to avoid possible cracks during the bending.

Just as with the cold processes, scrolling wrenches, forks, and other accessory tools are used to make the curves and twists. Tongs are often used to manipulate the heated steel.

Hot bending a square rod on a bending fork clamped in the blacksmith's vise, using appropriate tongs for safety

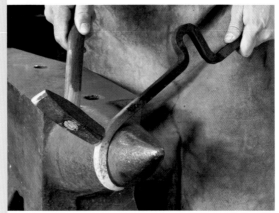

Curving the end of a piece on the round anvil horn

Using tongs to bend a strap into a spiral shape on a custom-made template

Detail of bends and curves created through heat: gate of the Güell Winery by Antoni Gaudí

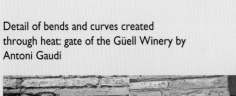

Hot twisting a rod in the blacksmith's vise with the help of a scrolling wrench. The rod was flattened in advance at the area to be twisted.

Basic techniques

Rolling

Rolling involves bending the steel onto itself to create spirals and scrolls. Hammer the heated end directly on the face of the anvil to make a small curve, and continue with successive heatings and hammer blows. It's also possible to create special templates to make several identical spirals. To bend the steel with a minimum of effort and to reduce marks left by the hammer, work the piece at a light orange color.

MAKING A SPIRAL ON THE ANVIL

This is one method of making a forged spiral on the end of a round rod.

1 Hammer a heated rod that has been tapered in advance on the round horn of the anvil to create a slight curve in the tapered end.

2 Hammer the curved steel on the face of the anvil in order to fold the end back onto itself.

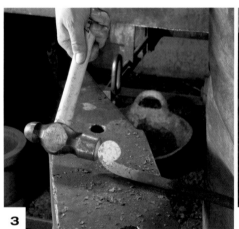

3 Heat the curved end again, and hammer on the face of the anvil in the direction of movement for the scroll.

4 Adjust the curves of the spiral with gentle hammer blows while the steel is still glowing.

The end of the rod rolled into a spiral

A twisted scroll

A twisted scroll is a spiral with the ends stretched out to increase its volume. The steps below illustrate how to make a spiral on the blacksmith's vise.

1 Bend over the previously tapered end of the rod gripped in the jaws of the blacksmith's vise.

2 Heat the rod again, hold the tapered end in the jaws of the blacksmith's vise, and begin curving the rod onto itself.

3 Use the hammer to adjust the spiral as it is being curved.

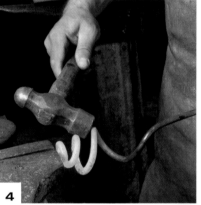

4 Once all of the turns of the spiral have been made, the rod is pulled out. While doing this, clamp the heated spiral in the vise by the tapered end. Pull on the part of the rod that is still straight, helping the process along with gentle hammer blows.

5 Use a steel bar to finish evening out the distances between turns. This step can be done cold if the thickness of the rod allows.

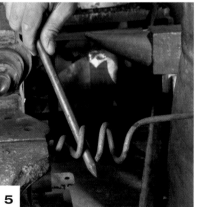

Sequence for creating and stretching out a twisted scroll

Right-angle bending

A right-angle bend involves creating a 90° bend in a piece of steel. The bend can be produced by hammering the heated steel directly on the anvil, as shown here, or by clamping it in the jaws of the blacksmith's vise and bending it with gentle hammer blows. Sometimes, to create the corner angle, it is necessary to first draw the metal out into a tenon or start with an upset in the area where the corner is to be bent.

Bending a drawn-out piece of steel on the side of the anvil's face

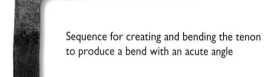

Sequence for creating and bending the tenon to produce a bend with an acute angle

1 Bending a heated piece of steel by hammering it on the side of the anvil face

2 Upsetting the angle on the face of the anvil

3 Adjusting the right-angle bend on the side of the anvil

Clips

Clips are a type of joint that involves rolling a piece of steel, usually a flat strap, around other pieces of steel to hold them together. It is necessary to know the unrolled length of the clip, which corresponds to the length of its neutral fiber. This equals the sum of the side lengths of the rectangle that results from combining the shapes to be joined with the length of the arcs in the clip. It's possible to make an approximate determination by measuring with a piece of string around the pieces of steel.

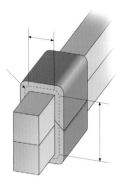
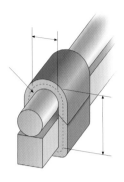

To calculate the length of the neutral fiber (the red line shown in the illustration), add together the lengths of the sides plus the lengths of the arcs to be enclosed by the strap.

1 The process of installing a clip. The previously bent and heated strap is placed around the two pieces of steel to be joined.

2 The strap is immediately bent around the two pieces of steel using gentle hammer blows. To avoid stretching the heated clip, don't hammer forcefully. When the steel cools, it contracts and holds the pieces together tightly.

Detail of the clip on a grille in Barcelona, Spain

Techniques for **Cutting and perforating**

This section includes the techniques for cutting off material or creating holes without the loss of steel. The material must be heated to the proper temperature so that it can be parted easily during forging operations. Anvil hardies and punches are necessary for the process. Although these are simple techniques to perform, they require lots of trial and error in order to gain experience. This, of course, is the only way to achieve high quality work.

Trimming

Trimming is done with hammer blows on a hardy secured in the anvil. Depending on the thickness of the stock, this can be done either hot or cold. The steel must be marked at the correct distance all the way around without completing the cut entirely. The end to be cut off is then hammered on the edge of the anvil face to separate it.

Hot trimming the end of a rod to make the head of a forged nail. The place where the cut was to be made was first marked out hot on the hardy.

1 The cold rod is struck with a hammer on the edge of the hardy. This is repeated on each of the four sides of the rod.

2 The end of the piece is then hammered on the edge of the anvil face until it separates from the rest of the rod.

1

The steps for trimming a rod properly

2

Ripping

Ripping involves dividing the end of the steel using a cut of a chosen length. This method does not separate material off the original shape, but instead opens it up.

The piece is heated to a uniform light orange color and placed on the hardy. Then it is carefully hammered on both sides, making sure not to damage the hammer on the cutting edge. This operation is always done hot, and reheating is done as many times as needed to achieve the complete split.

1 Start the split by hammering the strap on both sides while holding it on the hardy.

2 Apply vertical hammer blows over the cutting edge of the hardy to separate the two resulting sides and continue forging the piece.

This forged iron cross in Torroella de Montgrí, Spain, clearly shows the split ends that were subsequently tapered and bent.

Splitting

Splitting involves parting the steel in the center to create openings without removing any material from the stock. The bar is heated to dark yellow, at which point the steel is softest. It is placed over the hardy and hammered, turning it over to create the cut on two sides. Then the piece is struck vertically on the anvil to separate the resulting sides, and it is worked with the hammer on the conical beak to adjust the shape.

1 The glowing bar is hammered on the hardy, turning it over to cut on both sides.

2 The bar is struck vertically on the face of the anvil to separate the two sides.

3 The separated sides are forged on the conical beak of the anvil.

Sequence of steps in forging a split

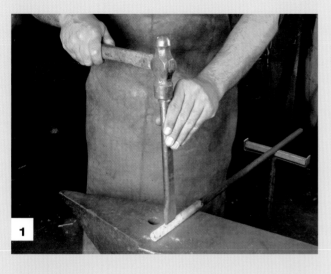

1 The area to be punched is marked with a cold chisel. It's helpful to support the piece with a work stand to free up your hands.

2 The hole is punched on the round hole in the anvil face, on both sides of the rod.

3 The pierced area is upset to preserve the thickness and to limit the stretching caused by inserting the punch.

Punching

Punching is a combination of several forging techniques. First, a split is created, then it's upset, and finally it's widened to create a hole of a specific diameter.

The operation is done over the round hole in the anvil, turning the rod to start the punch on both sides. The final hole is achieved after heating the rod several times to gradually and evenly stretch it.

Sequence of the punching process

Detail of a grille showing a joint in the bars created by punching.

Special **cutting techniques**

These steel cutting methods are affordable for small workshops. An oxygen cutting torch allows not only for cutting, but also can be used for heating metal. A mixture of acetylene and oxygen in the torch combusts to produce heat.

Plasma cutting allows for fast cutting of any metal, although it does require a certain amount of workshop infrastructure.

Oxygen cutting

Oxygen cutting is designed for ferrous metals. If the temperature of the steel is raised to above 1652°F (900°C) and a jet of pure oxygen is applied to it, the carbon steel combusts. The flame serves to raise the steel piece to the

ignition temperature and to clean the surface of slag, oxides, and leftover paint. The jet of oxygen is then applied through the same torch to burn the metal when it reaches the ignition temperature and to remove the slag formation, producing a narrow groove.

1 The pre-heating flame is applied to the start point of the cut for a certain amount of time; the time varies depending on the thickness of the metal to be cut.

2 The jet of extra oxygen is applied to produce the combustion of the steel, and the torch is moved in the direction of the cut.

3 A constant distance is maintained between the nozzle of the torch and the steel being cut to guarantee a uniform temperature throughout the process.

Ares, *Petita història d'un cilindre*, 2005. 19.6 x 23.6 x 3.1 in (50 x 60 x 8 cm). Detail of pieces produced by the oxygen-cutting process after the edges are ground.

1 The nozzle of the torch is placed off the piece to begin cutting. It's necessary to drill holes as starting places for the cuts for open work.

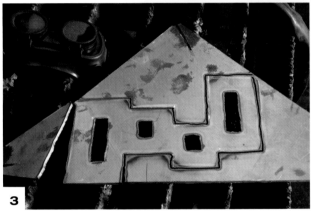

Plasma cutting

This cut is created with a stream of plasma produced when a gas (or mixture of gases) passes through a narrow nozzle and is then ionized with energy supplied to it by a strangulated electric arc. In addition to the thermal action, which melts the metal with temperatures reaching 36,032°F (20,000°C), the plasma stream exerts a mechanical action as it moves at a high speed of about 3,280 feet per second (1,000 m/s), continually freeing up the melted material. Air, a mix of nitrogen and oxygen, is most commonly used for these operations. In contrast to the oxygen-cutting process, plasma cutting makes it possible to cut any metal that conducts electricity.

2 The ground clamp is placed on the workbench to produce the stream of plasma. The bench must allow the molten metal to flow out from beneath the piece.

3 Due to the pressure of the plasma stream the melted metal is continually liberated during the process. This produces a clean cut in the metal, with minimal heating.

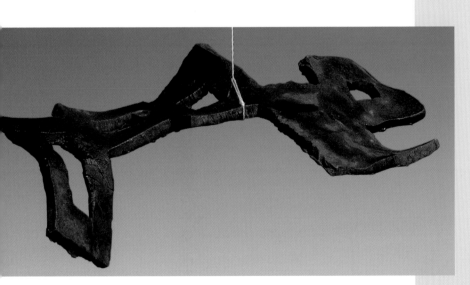

Ares, *De l'aire*, 2006. 10.6 x 6.7 x 3.9 in (27 x 17 x 10 cm). Produced by plasma cutting and forging.

Basic techniques

Heat **treatments**

In this process, metal is heated to a certain temperature, kept there for a time, and cooled at a specific rate. The results depend on the temperature reached and the speed of the cooling. This process of heating and cooling changes the metal's structure and its mechanical properties. This section explains heat treatments that do not change the composition of the metal aggregates. These are annealing, hardening, and tempering.

Annealing

This treatment restores qualities (including ductility) to the steel that were lost through mechanical operations such as cold hammering. The piece is heated to a specific temperature between about 392° and 1292°F (200° and 700°C), held there for a certain time, and slowly air-cooled.

Hardening

This heat treatment increases the toughness and hardness of the metal, but also its breakability. Hardening involves heating the steel evenly to a certain temperature, between about 1382° and 1472°F (750° and 800°C), then quickly quenching it in a bath, usually water at room temperature.

The piece is plunged into the water in a very deliberate manner. The container of liquid must be large enough to accommodate the entire piece as it is moved around. It's best to put it into the bath along its vertical axis. Its shape will determine how best to stir it around in the liquid. This treatment is very commonly used for making steel tools with a carbon content of under 0.9%.

Tempering

Tempering helps to eliminate the tensions produced by sudden expansions and contractions when the steel is hardened. A hardened piece is reheated to a relatively low temperature, between 212° and 608°F (100° and 320°C), and then cooled. In the tempering process the piece loses some of its hardness, but it gains resilience.

Forged tools that have been hardened and tempered for working stone and marble

Hardening a tool for cutting stone

A gradine is made from a round rod of low-carbon steel (less than 0.9%) that is forged and hardened by the blacksmith to be used for working stone. The process of heat-treating tools requires lots of observation and practice.

1 The steel rod is heated and pounded out on the face of the anvil. This involves broadening and tapering the end of the rod that subsequently will be hardened.

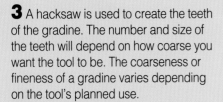

2 Once the end is forged, the electric bench grinder is used to create a cutting edge, using a fine-grit wheel.

3 A hacksaw is used to create the teeth of the gradine. The number and size of the teeth will depend on how coarse you want the tool to be. The coarseness or fineness of a gradine varies depending on the tool's planned use.

4 Heat the portion that is to be hardened. Since the steel must be heated slowly and uniformly, keep turning it over in the forge. When the piece reaches cherry red, it is at the hardening temperature, between 1310° and 1562°F (710° and 850°C). Plunge the rod into the quench (water) vertically and in a straight line. Note how the hardening colors climb up the steel. (Refer back to the chart on page 20.)

WELDING TECHNIQUES

This section presents two of the most common welding processes used in smaller workshops. Mastering welding techniques is only possible with intensive practice—a thousand trials and errors until you consistently achieve good results. When working with welding equipment, you must always protect yourself against ultraviolet ray emission, red-hot sparks, and burns.

Arc welding

In arc welding with a coated rod, the metal is fused through heat generated by an electric arc, also referred to as a voltaic arc, set up between the end of the coated electrode and the metal base. This process requires great skill and practice to control all the parameters. On the other hand, arc welding with a coated road is one of the most accessible techniques because the necessary equipment is widely available.

The coated rod

The electrode is responsible for setting up the voltaic arc, protecting the fusion bath, and supplying the material for the joint. It is made from a metal rod coated with a material composed of various chemical substances. The most common type is made from rutile, a form of titanium dioxide. For welding steel, a metal rod made from different ferrous materials is used for joining the edges of pieces by fusion.

The coating stabilizes the voltaic arc and creates gases that protect the fusion bath. It also forms a slag that covers the bead of weld to prevent sudden cooling, which could cause cracks, and it simultaneously keeps this red-hot area from contacting the oxygen and nitrogen in the air. That would cause the bead to oxidize and would create pores inside it, thereby weakening it.

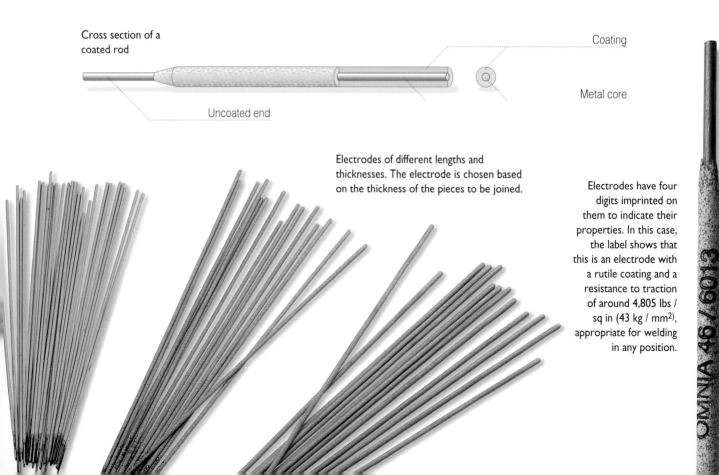

Cross section of a coated rod

Coating

Metal core

Uncoated end

Electrodes of different lengths and thicknesses. The electrode is chosen based on the thickness of the pieces to be joined.

Electrodes have four digits imprinted on them to indicate their properties. In this case, the label shows that this is an electrode with a rutile coating and a resistance to traction of around 4,805 lbs / sq in (43 kg / mm^2), appropriate for welding in any position.

OMNIA 46 / 6013

Basic technique

Placing the electrode into the clamp. Protective gloves are used to guard against possible electric discharge.

ELECTRODE PLACEMENT

The uncoated part of the electrode is placed in the clamp of the electrode holder. Because there is still tension (voltage) in the clamp even when you are not welding, wear protective gloves to avoid the risk of electric shock.

STARTING THE WELD

To begin, place the end of the electrode held in the clamp in contact with the pieces of steel to be welded, which are held by the ground clamp. This is called priming. The voltaic arc is established, the fusion bath is created, and this forms the welding bead that joins the pieces. A side-to-side motion is applied to the electrode while welding.

To begin welding, the electrode is tapped lightly at the point where the weld is to begin, and is immediately pulled back slightly to establish the electric arc.

SAFETY

Throughout the welding process, wear a leather apron and leather gloves to protect against flying sparks, and wear a mask with a non-actinic glass for protection from the ultraviolet rays produced by the fusion bath.

Proper eye protection, an apron, and gloves help ensure safety during welding.

Diagram showing the movements of the electrode: circular (A), semicircular (B), zigzag (C), and interlaced (D)

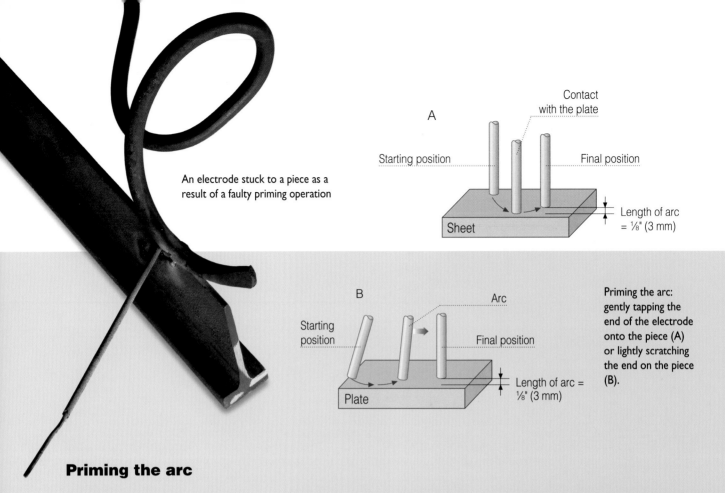

An electrode stuck to a piece as a result of a faulty priming operation

A

Starting position

Contact with the plate

Final position

Length of arc = ⅛" (3 mm)

Sheet

B

Starting position

Arc

Final position

Length of arc = ⅛" (3 mm)

Plate

Priming the arc: gently tapping the end of the electrode onto the piece (A) or lightly scratching the end on the piece (B).

Priming the arc

Priming involves lightly tapping the tip of the electrode onto the piece to be welded (connected to the ground clamp) and immediately pulling it back a distance equal to the diameter of the electrode. This establishes the voltaic arc and begins the weld. No matter what technique is used to prime, it is important to immediately lift the electrode to the proper height to set up the arc. Otherwise, the electrode will stick to the piece and will quickly get red hot from the electric current passing through it. If this happens, quickly release it from the electrode clamp or detach it from the piece. Because of the intense light, these steps must be done without removing the face shield from the eyes. After the voltaic arc is established, proceed with the proper speed,

intensity, electrode angle, and arc length. Be sure not to go too slowly and waste welding rod by creating too coarse a bead, or too fast, which may prevent an even, uniform bead. A sense for the correct speed

will develop through practice. The characteristic sputtering of the welding process will indicate that everything is correct.

Diagram of the welding process with coated rod once the arc is primed

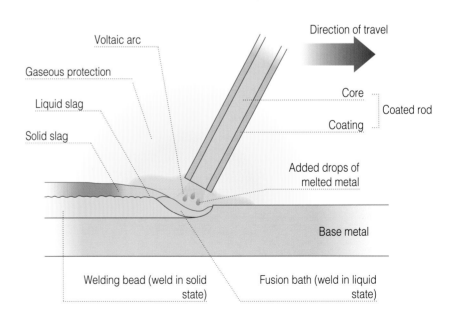

Voltaic arc

Gaseous protection

Liquid slag

Solid slag

Direction of travel

Core

Coated rod

Coating

Added drops of melted metal

Base metal

Welding bead (weld in solid state)

Fusion bath (weld in liquid state)

Welding parameters

Several parameters dictate the process of arc welding with coated rod.

The **diameter** of the electrode is chosen with respect to the thickness of the material, the joint, and the position of the weld. Generally, thin electrodes should be used on thin pieces and on the initial spots for joints.

The **intensity** of the current in the welding process affects the degree of penetration of the weld into the base metal. The greater the intensity, the greater the resulting penetration. When welding at an angle, the intensity must be much greater than in other positions to ensure that the fusion bath penetrates adequately.

The **length of the voltaic arc** also influences the quality of the weld. In general, it should be the same as the diameter of the electrode.

The **speed of travel** during the welding process must allow the arc to go slightly ahead of the fusion bath. This will prevent overheating and will produce narrow beads of weld that cool quickly.

The **angle** is defined in two ways: by the longitudinal inclination formed by the electrode and bead of weld, and by the lateral inclination formed by the electrode and the pieces to be welded together. If the angle isn't correct, slag may get inside the bead.

Crystallized slag typical of arc welding with coated rod

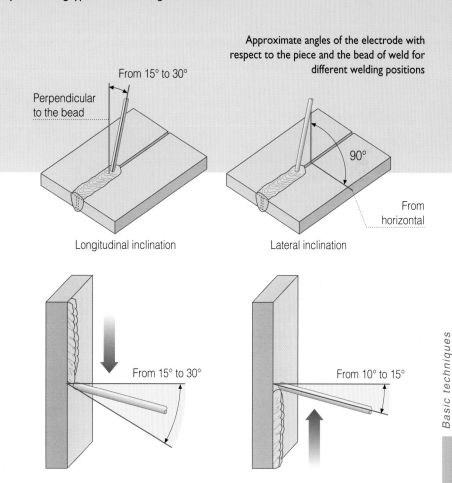

Approximate angles of the electrode with respect to the piece and the bead of weld for different welding positions

From 15° to 30°

Perpendicular to the bead

Longitudinal inclination

90°

From horizontal

Lateral inclination

From 15° to 30°

From 10° to 15°

Basic techniques

Oxyacetylene **welding**

In the oxy-gas welding process, a flame from the combustion of a gas and oxygen serves as the energy source for the fusion. The most commonly used gases are oxygen and acetylene. This combination generates a temperature of almost 5800°F (3200°C). The flame heats the pieces to be joined until the contact areas melt, producing a bead of weld.

Oxyacetylene welding can be done with or without adding extra metal, called filler. When filler is used, it is applied in the form of metal rods, usually of the same type as the base metal. This is well suited for joining thin pieces and for strong welds in copper or brass.

SETTING UP THE EQUIPMENT

The manometers for different gases shouldn't be switched between different tanks, and all safety instructions provided by the gas tank distributors should be carefully followed. The torch must have a built-in check valve to keep the flame from backing up.

The manometers must be regulated with respect to the diameter of the nozzle. In general, oxygen is adjusted to a working pressure of around 4.4 to 6.6 pounds per square inch (2 to 3 kg/cm^2); the pressure for acetylene is between about 1¾ and 3 ounces per square inch (50 and 100 g/cm^2).

Protective goggles and face shield are necessary to reduce the effect of the intense light during the welding operation

Welding torch and interchangeable nozzles; different diameter nozzles are used for different thicknesses of pieces to be welded. Note the check valve for the flame, located between the body of the torch and the hoses.

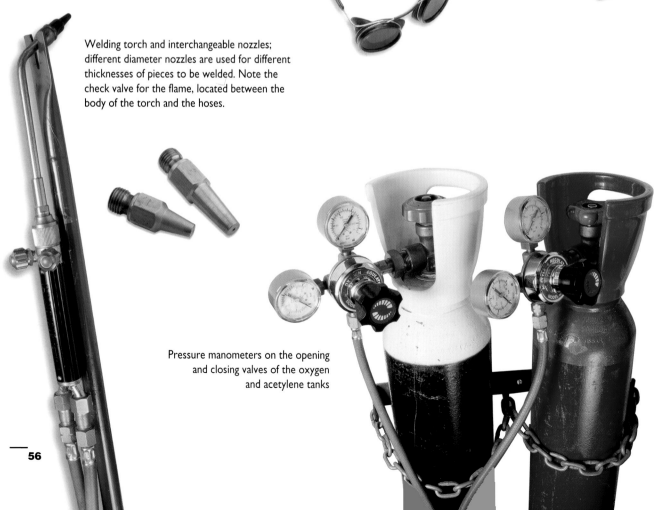

Pressure manometers on the opening and closing valves of the oxygen and acetylene tanks

THE FLAME

There are two distinct parts of the flame. One is the intensely dazzling white cone, where the combustion of the acetylene and the oxygen takes place; the other is the plume that surrounds the cone and protects the fusion bath. There is a third area that is not perceptible to the naked eye known as the work area; it is located immediately beyond the cone, and it is the area of highest temperature.

LIGHTING THE TORCH

Open the gas valves on the tank of acetylene. The acetylene is lit with a lighter or a match. The oxygen valve is opened slowly to produce a neutral flame of an equal amount of acetylene and oxygen. Remember to wear protective eye wear against the intense light. To turn off the torch, close the acetylene valve first, then the oxygen valve.

TECHNIQUE

The cone of the flame is brought to within ³⁄₃₂ to ⅛ inch (2 to 3 mm) of the base piece to cause the fusion of the two edges. A subtle turning of the cone at the instant the edges flow causes them to fuse. The resulting drop is pushed with the pressure from the cone at a 45° angle with respect to the welding bead, to weld without adding metal.

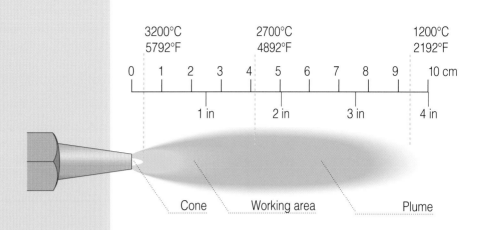

Parts of the flame and the relation between the temperatures and distances from the cone

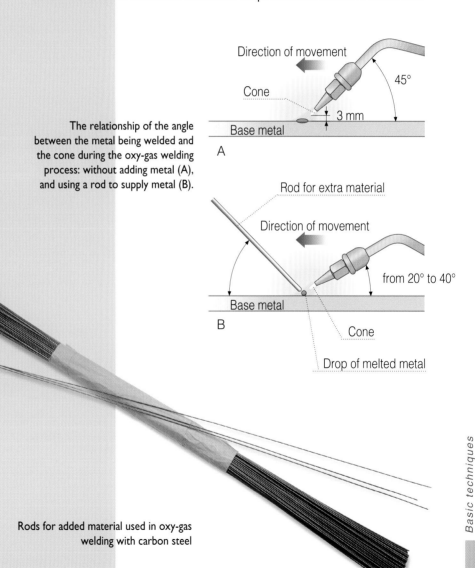

The relationship of the angle between the metal being welded and the cone during the oxy-gas welding process: without adding metal (A), and using a rod to supply metal (B).

Rods for added material used in oxy-gas welding with carbon steel

Step
by step

Now that you have practiced with the Basic techniques, materials, and tools of forging, here is a series of practical exercises for making forged objects. These step-by-step projects help to describe the techniques explained in the previous chapters.

Although these five exercises are not meant to be imitated exactly, they serve as a useful guide for better understanding the different techniques, and for trying your own projects.

Cold-forged
trivet

Making a simple but practical table trivet is a good introduction to the techniques of cold forging. A sheet of ³/₃₂-inch-thick (2-mm-thick) steel, left over from another project, will be cut into strips. These will be twisted without heating and will be arranged in a grid pattern. Self-tapping screws will be used to join the strips.

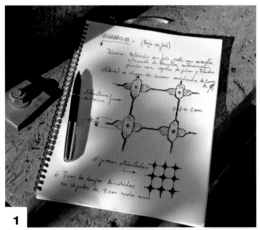

1 A sketch of the trivet is done in advance; planning ahead facilitates the work.

2 Use a scratch awl and a metal rule to mark strips about ⁹/₁₆ inch (1.5 cm) wide on the steel.

3 Use a shear to cut out the strips along the marked lines. Be sure to wear leather gloves.

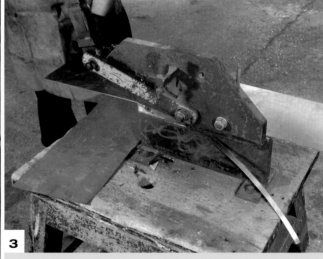

4 Straighten the strips by hammering them on the anvil.

5 File the strips to remove the sharp edges.

6 To identify where the twists will be made, mark the strips into ¾-inch (2-cm) segments with a square and a scratch awl.

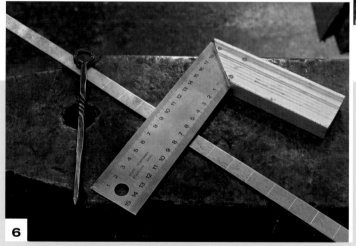

7 Divide the strips into segments of about 11¾ inches (30 cm) in length, and use the bench grinder to round the ends.

Step by step

8 Use a scrolling wrench to start the twist in each section of the strip held in the blacksmith's vise. The previously marked divisions indicate the distances between twists.

9 Make the twists in the same direction until the entire strip is completed.

10 The strips are the same length, thanks to the previous scribing of the twists to equally space them.

11 Use a round file to smooth any imperfections.

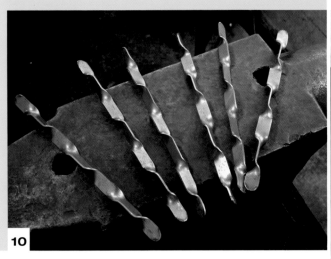

"The strips are the same length, thanks to the previous scribing of the twists to equally space them."

12 On the square beak of the anvil, use a center punch to mark the placement of the holes where the strips will be joined together.

13 Drill these holes with an electric drill, using a bit slightly smaller in diameter than the self-tapping screw that will be used to connect the strips.

12

13

14

14 Use a fine file to remove the burrs produced by drilling.

15 Follow the method of one turn ahead and a half-turn back to cut the threads properly in the strip and install the screws.

16 Screw the strips together, keeping the screw slots parallel to each other to create a harmonious look.

15

16

17 Use an angle grinder to cut away the remainder of the screw projecting through the sheet metal.

18 With a steel brush chucked up in an electric drill, even out the surface of the trivet.

17

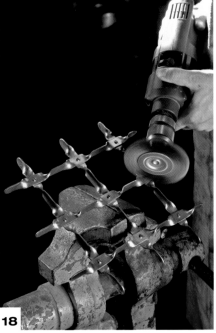

18

19 Apply a transparent spray varnish to protect the metal from rust. Once it's dry, apply a coat of wax to cut down on the shine of the metal.

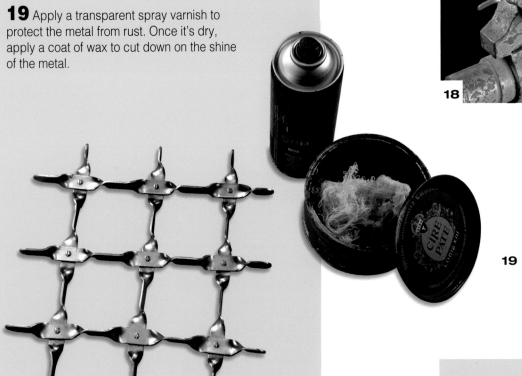

19

The finished trivet

Door
handle

Making this door handle requires the basic techniques of hot twisting and rolling. The door handle will consist of a twisted central grip and two spirals at the ends. The material is a rod about ⅜ inch (10 mm) in diameter and 3¼ feet (100 cm) long. Screws will also be made to attach it to the door.

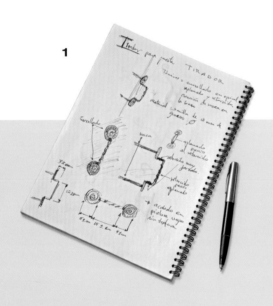

1 Make a few sketches of the handle, and note down the measurements and techniques you plan to use.

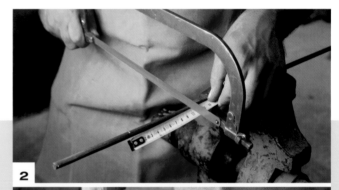

2 Using a hacksaw, divide the steel rod into seven equal lengths.

3 Place the heated rod in the blacksmith's vise and bend the two ends to 90° angles with a hammer.

4 Flatten the center of the rod by hammering it on the face of the anvil.

5 Turn the rod as it flattens to shape it in cross-section.

6 Start the bend for the spiral by supporting the bent part of the rod in the blacksmith's vise and striking it with the hammer.

7 Strike the heated rod with the hammer to roll up the spiral, working continually at a bright red heat.

8 Occasionally, hammer the spiral gently to even it out.

9 The first spiral is finished when you get to the flattened part of the handle.

10 Clamp the other end of the rod in the blacksmith's vise to make the second spiral.

11 Make the second spiral using the same process as for the first one.

10

"Strike the heated rod with the hammer to roll up the spiral, working continually at a bright red heat."

11

12

12 Use water to cool down the material where it is held.

13 The second spiral is completed when you reach the flat section of the door handle.

13

14

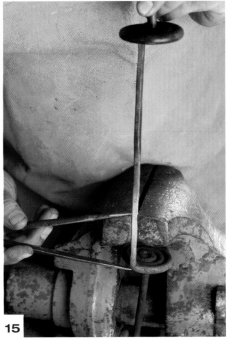

14 To make the grip portion, bend the rod perpendicular to the spirals.

15 The compass is used to quickly mark the spot where the grip is to be bent.

15

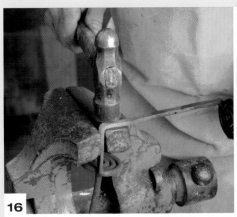

16

17

18

16 With a hammer, bend a section of the grip at red heat on the blacksmith's vise.

17 The other end of the grip is bent using the same technique so that the spirals orient in the same direction.

18 At this stage of the process, the rod looks like this.

19 Place the heated door handle in the blacksmith's vise, and clamp it by a spiral. Use flat-jaw tongs to move the spirals to the outside of the door handle.

20 Do this for each end of the handle.

21 Using the tongs, gently twist the handle.

22 Use a scrolling wrench to accentuate the twist on the sides of the spirals.

23 Cut several threads on the ends of the spirals to attach the handle to the door.

An exterior enamel provides the perfect finish for the door handle.

Fireplace
tongs

These fireplace tongs will involve the basic forging techniques of drawing out a tenon, bending, hammering out, and riveting. The tongs will be made from two straps of steel, each ⅛" thick, ³⁄₁₆" wide, and 27½" long (3 mm x 20 mm x 70 cm), and will have loop ends. The upper loop is used for holding and operating the jaws while the lower part comes in contact with the heat and houses the central hinge.

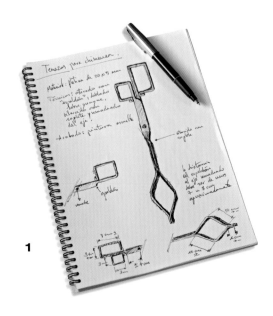

1

2

3

1 Planning comes first: design the tongs on a piece of paper.

2 Make the loop ends first. Mark the straps where the tenon will start.

3 Heat the steel and strike it on the edge of the anvil to mark out the start of the piece to be drawn out.

4

5

6

4 On the anvil, vigorously hammer the strap at bright red heat to draw it out from the marked point.

5 Draw out the whole length to make the loop ends of the tongs.

6 While drawing out the steel, it's a good idea to turn it over to keep the surfaces even.

7

7 Form the loop for one side of the tongs in the tang and the drawn-out tenon.

8 Put a finishing curl into the end of the loop. First, fold over the tip of the drawn-out section of strap on the edge of the anvil face.

9 Roll the strap back onto itself by lightly hammering on the previously bent tip.

10 On the flat surface of the anvil, strike the top of the spiral so that it will have a square outline.

11 Continue shaping the square inside the spiral by hammering it on the corner of the anvil face.

"While stretching, turn the strap to equalize the surfaces."

12 Start the loop for the tongs on the square beak of the anvil. Use a compass to mark the spot where the loop will be bent.

13 Turn over the loop for the tongs to produce a 90° angle.

14 Complete the loop by closing up the square. Proceed in the same way for the other arm of the tongs, following the design.

15 The completed loops.

16 On the edge of the anvil, fashion the arms of the tongs from the jaws by marking the ends of each arm.

17 Upset the tips of the jaws to achieve a rounded shape.

18 To produce a triangular curve in the arm, it is heated and bent with the help of a fork inserted in the square anvil hole.

19 Using the first arm as a guide, bend the second one.

"The head of the rivet is finished off in a heading tool."

20 Make the pivot for the tongs in the form of a rivet. Hold a piece of ¼-inch rod (6 mm) in the jaws of the vise. Heat the rod with a torch while it is upset by hammering.

21 Finish off the head of the rivet in a heading tool.

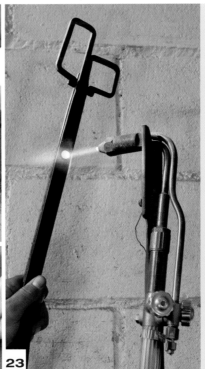

22 Use an electric drill with a drill bit smaller than half the width of the arm to drill the hole for the pivot. The arms of the tongs will pivot on this point.

23 Put the rivet into the hole, cut it to the needed length, and heat it with a torch.

24 Immediately upon heating, upset the pivot to shape the head of the pin. Perform the operation on a base plate installed in the square hole of the anvil.

25

25 To finish the tongs the arms are bent 90° to the pivot point for proper functioning. They are heated at the point for the twist, using a torch to localize the heat more effectively.

26 Twist the tongs using a scrolling wrench while they are clamped in the blacksmith's vise.

26

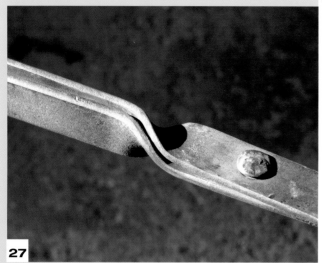

27

27 Detail of the twisted area.

Fireplace tongs finished with flat
black enamel

Grille with
scrollwork

This project is a scrollwork grille with two symmetrical sides. Templates are constructed cold for each scroll, and the steel straps are forged on them. The two scrolls will be welded together, framed inside straps of steel, and connected to the frame with clamps. This project's design is modeled after a forged railing on an interior patio of the Can Ballester in Sitges, Spain, a restoration project by Valerià Cortés.

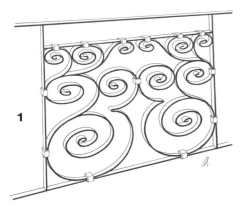

1 Make a sketch of the scrolls in the grill to create a general guide for the final result.

2 On a sheet of plywood, draw the scrolls to size in the desired shape.

3 To make the grille, first construct a metal template with the shape of the scrolls. Using a bending fork and bending hook, curve a thin strap of steel.

4 At the same time, check the curves of the strap against the drawing on the plywood.

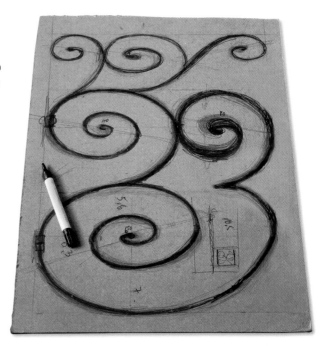

5 and **6** Weld the resulting strips to a steel plate that has a piece of steel welded to the bottom that will fit into the square hole in the anvil. Weld the strap steel to the sheet in two phases to allow forging every curve of the scroll.

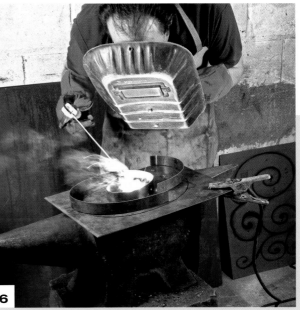

7 Using a length of rope placed over the line drawn on the wood, determine the approximate length of the steel strapping needed to make the scroll.

8 Stretch the rope along the steel strapping and use a piece of chalk to mark off the distance to cut. Because two scrolls will be made on the same template at the same time, perform this step (and those below) on two different pieces of strapping at the same time.

9 To make the first scroll, heat the tip of the strapping and bend it slightly on the face of the anvil.

10 Heat the strapping again and hook it into the start of the template; forge the first curve by exerting pressure on the template.

11 As the curves of the scroll are forged, the templates are welded to the steel plate.

12 To make sure that the material isn't heated to the melting point, remove it from the fire at the proper temperature, as shown by the color of the steel in the photo.

"As the curves of the scroll are forged, the templates are welded to the steel plate."

13 Heat the strapping in the forge and work it on the anvil, following the template for each curve of the scroll.

14 Use a flat piece of steel as a lever to remove the forged strapping from the template.

15 Make each curve of the scroll in duplicate, since the grille will have two symmetrical parts.

16 After finishing every scroll, use the proper tongs to hold the red-hot steel, and hammer it to conform to the template.

17 Complete the last curve of the scroll by pressing the strapping against the template with the tongs.

15

Step by step

18 Place the spiral onto the plywood drawing. Use chalk to mark the point where the extra strapping will be cut off.

19 Use a cutting wheel to cut off the spiral at the chalk mark.

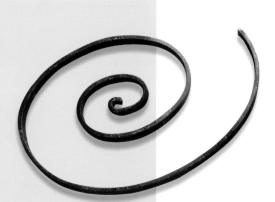

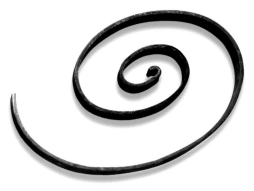

20 The finished scroll pieces made with the same template, after the extra material has been cut off.

21 Using a square, attach steel rods to a sheet of steel to temporarily arrange the scrolls.

"Make sure that the material does not heat up to the melting point."

22 Using an angle grinder, make the slots where the scrolls fit together.

23 Close-up of how the scrolls fit together. The shiny area that was ground is the place where the scrolls will be welded together. Note the slight bevel on the edge of the strapping to make room for the weld.

24 Continue grinding each scroll individually, and fit them together on the temporary plate.

25 The finished scrolls and frame after fitting them together on the steel plate according to the initial drawing.

Step by step

85

26 Hold the spirals together by welding with coated rod. Take particular care in priming the arc on the scrolls to avoid any shift in their alignment.

27 The bead of weld at the joint between two scrolls. Note the crystallized slag that is typical of welding with coated rod.

28 Go over the welds with an angle grinder to even out the surfaces.

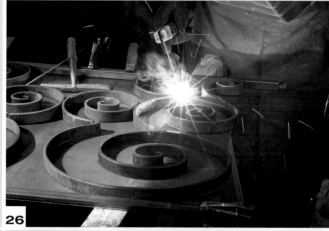

26

27

28

29

29 The grille after the welds at the joints between scrolls have been ground.

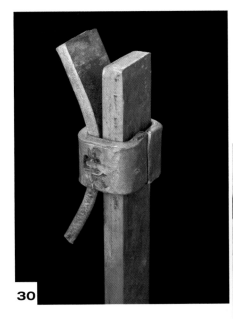

30 Next, make the clamps to hold the scrolls to the frame. Once you determine the length of the neutral fiber, make a sample and try it out.

31 Heat the strapping with a torch on a die block made to dimensions for fabricating numerous clamps of the same size.

" When it cools, the clamp shrinks, thereby increasing its grip and strength. "

32 and **33** To create the U shape, hammer the heated clamp into the die block with a bar made for the purpose.

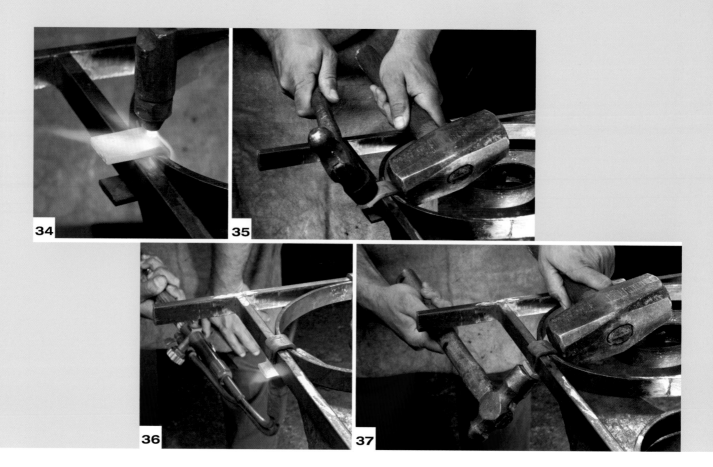

34 Place the bent clamp at a point on the frame where it touches a curve of the scroll. Heat it with the torch. The frame that will hold the scrolls by means of the clamps has been constructed.

35 Start bending and closing the heated part of the clamp by hammering it; a sledgehammer held behind it will keep the scroll from separating.

36 and **37** Heat the rest of the clamp and continue working as in the previous step.

38 The clamp shrinks as it cools, increasing its grip and strength.

Scrollwork grille on the patio of the Can Ballester in Sitges, Spain. It was finished with a coat of reddish antirust paint to simulate rusted iron.

Wall
candelabrum

Our next project is to create a hanging candelabrum. It consists of three parts: the bracket from which the candelabrum hangs, the candelabrum itself, and a connector that holds it to the bracket. We will use the techniques of splitting, perforating, bending, flattening, tapering, and cutting while combining traditional techniques with modern processes. For example, the flame of a torch will be used to apply heat to a specific spot, and we'll use an angle grinder to cut a groove.

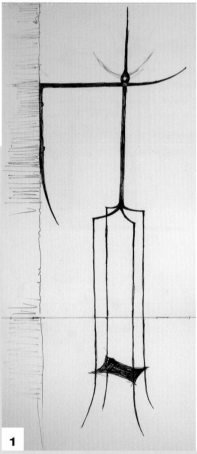

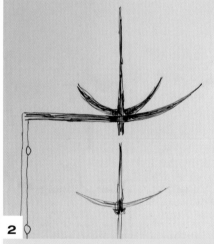

1 and **2** Plan the candelabrum project with some sketches. The sketches will serve as a guide for the piece you want to create.

3 Start the split by marking the center of the heated rod with a cold chisel and a hammer on the face of the anvil.

4 When the rod is at yellow-red heat, hammer it on a hardy in the anvil, and split it in the middle.

5 Next, strike the tip of the heated bar vertically on the anvil to open the two sides of the split.

6 Place the opening on the rounded anvil beak, and hammer the tip of the rod to produce a rounded shape.

7 Use gentle hammer blows to refine the shape on the rounded beak.

"Gentle hammer blows are used to refine the shape."

8 Heat the end of the split rod and taper it by hammering forcefully on the face of the anvil.

9 and **10** Keep turning the rod while hammering to create a uniform point on the tapered end.

11 Measure and mark the precise spot on the rod to bend to a 90° angle. When the rod reaches a red heat, use a bending fork on the anvil to create the bracket shape.

12 After heating, slightly taper and curve the opposite end.

13 Flatten out the areas for the screws that will attach the bracket to the wall. Heat the selected area, and hammer forcefully on the rounded beak.

14 Once the flats are in place, straighten the rod by heating and hammering it on the face of the anvil.

15 Detail of the flattened areas where the bracket will be attached to the wall.

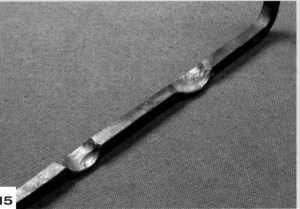

16 The completed bracket.

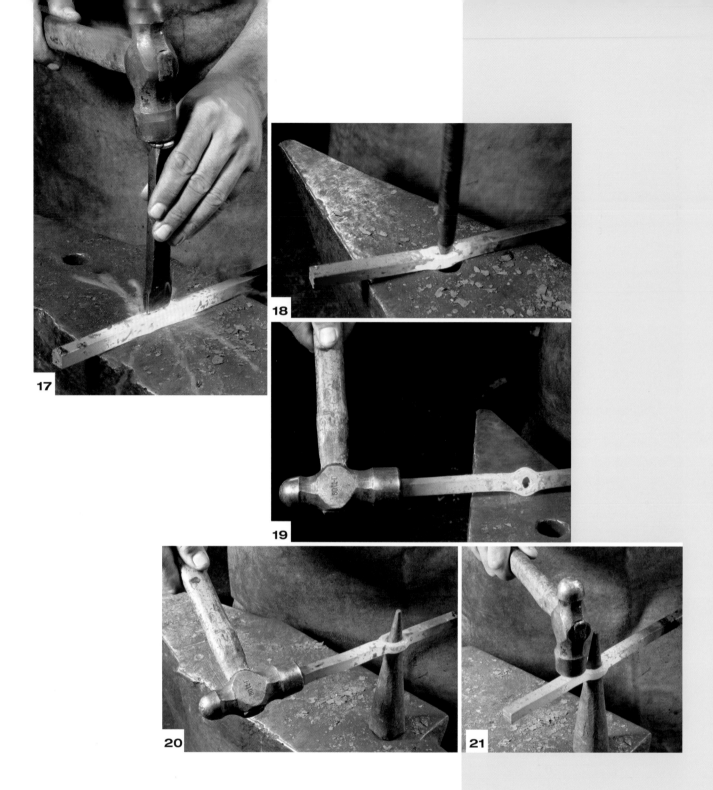

17 To make the main part of the candelabrum, punch a hole in the square rod when it has been heated to yellow-red. Mark the area first.

18 Place the rod over the round hole in the anvil face, and make a hole by striking the spot with a punch.

19 As the hole is punched, upset the area in order to counteract the stretching produced by the punching operation.

20 and **21** To produce a uniform hole, adjust it on an anvil tool so that you can hammer in all directions.

22 Draw out the end of the rod near the hole to create a tapered point on the candelabrum

23 Use a hardy with a rounded end to mark the place on the rod where it will be hammered out.

24 To hammer out the rod, strike it on a flat stake held in the anvil.

22

23

24

"As the hole is punched, it is necessary to upset the area."

25 The inner part before hammering out. Use a cutting wheel to make some grooves in the part of the rod that is not hammered out.

25

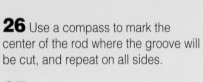

26 Use a compass to mark the center of the rod where the groove will be cut, and repeat on all sides.

27 Go over the lines marked with the compass with an angle grinder and a thin cutting wheel to make a groove sufficiently straight and deep to keep the disk from running off course.

28 Produce the cut with several passes of the grinder on the sides of the rod.

29 Heat the resulting ends of the groove, and then separate and bend them to a 90° angle to the main axis.

30

31

32

30 Once the sides are separated, go over them with a file to remove the rough edges produced by the grinder.

31 Heat and taper the ends on the face of the anvil.

32 Measure from the center and heat with a torch. Next, bend at the heated area using flat-jawed tongs.

"Go over the lines marked with the compass to make a sufficiently straight groove."

33 The central part of the candelabrum.

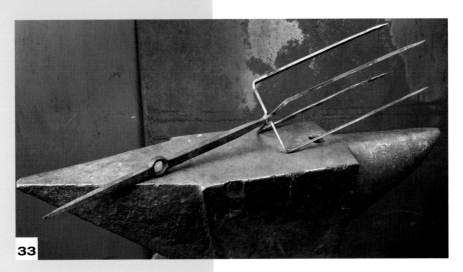

33

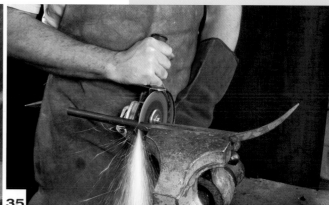

34 To begin the connector that will hold the candelabrum so it hangs from the bracket, first taper the end of an appropriately sized rod.

35 Clamp the rod in the jaws of the blacksmith's vise and cut off the excess from the opposite end with a cutting wheel.

36 Hold the rod with tongs, to avoid burning your hands, and taper the opposite end.

37 and **38** Heat the ends, and use tongs and a hammer to slightly curve them on the rounded anvil beak.

39

39 Use a plasma cutter to cut out a piece of sheet steel. To produce a straight cut, use a piece of angle iron as a guide for resting the plasma torch.

40 Heat the corners of the sheet steel, and curve them by striking them on the rounded beak of the anvil.

41 Use an electric drill to make holes in the curved ends, holding the sheet in the jaws of the blacksmith's vise.

42 Detail of the steel plate that will hold the candle.

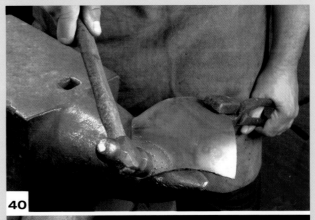

40

41

42

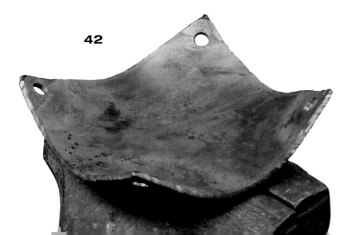

Step by step

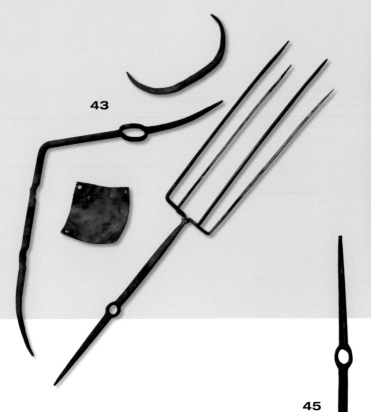

43

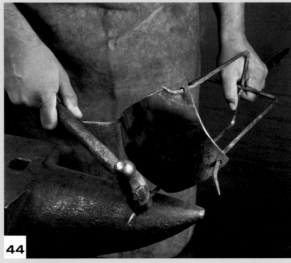

44

45

46

43 Breakdown of the pieces: bracket, holder, base plate for the candle, and the main body of the candelabrum.

44 The main body sides are fit into the holes in the base plate and bent cold by gently hammering on the rounded anvil beak.

45 The assembled main part of the candelabrum.

46 Detail of the junction involving the holder, the main body, and the bracket.

The candelabrum installed on a wall. The finish is a graphite patina.

Door
knocker

The following exercise shows how to make a door knocker shaped like a reptile. A dragon? A lizard? Whatever you reveal in your door knocker, the project involves producing the complex shape from a single section of round steel rod. Nails and a base plate are forged for attaching the knocker to the door. The pieces necessary for its movement are forged as well. Basic techniques are combined with modern ones such as oxyacetylene cutting.

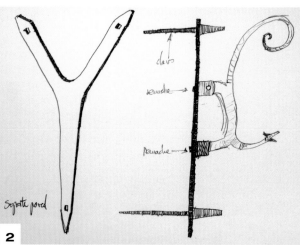

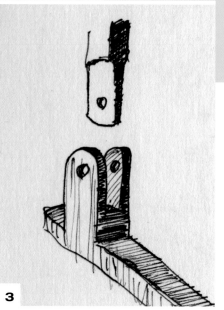

1 Sketch the main body of the door knocker, the lizard, which will be hammered out from a single piece of steel.

2 Sketch the base plate for the knocker and how it attaches to the door.

3 Detail of how to attach the knocker to the base and allow its operating movement.

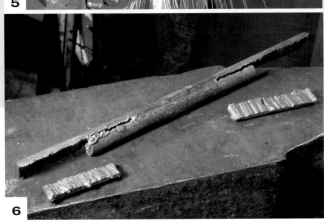

4 Use white chalk to mark the areas of the rod where the oxygen cutting will be done. The parts with the chalk will be cut off entirely.

5 Use the oxygen torch to cut off the marked areas; use appropriate goggles as protection against the intense light.

6 The piece after cutting, along with the discarded pieces that were previously colored white. The cuts in the rod mark out the various parts of the lizard: legs, tail, neck, and head.

7 Spread the cuts in the end by heating the rod and striking it along the cut line on a tall hardy.

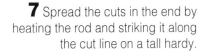

8 and **9** Hammer and taper the part that will become the lizard's tail. The long part previously spread is heated and hammered forcefully on the face of the anvil. Use appropriate tongs to hold the piece securely.

10 Hammer out the opposite end, which will form the neck of the lizard. Prior to this the steel has been spread in the area of the torch cut, as in step 7.

11 Heat the end and clamp the piece by the neck in the blacksmith's vise. Upset the end to compact the metal and create the head of the animal.

12 Reheat and slightly taper the upset area to shape the lizard's head.

13

14

13 and **14** Heat the part that corresponds to the front legs. Flatten it and upset the metal simultaneously to produce a rectangular shape.

15 Hammer out the part that corresponds to the hind legs. This is the area of the knocker that will be moved to tap against the base attached to the door.

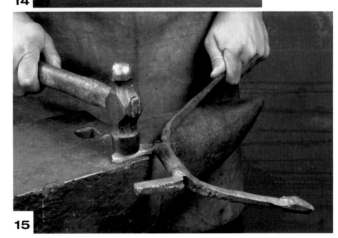

15

16

"Use appropriate tongs to hold the piece securely."

16 The finished lizard, consisting of one single piece without any joining.

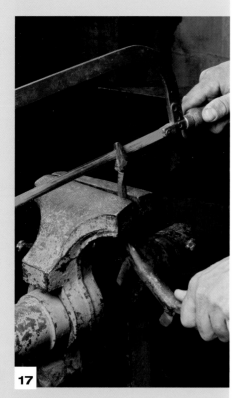

17

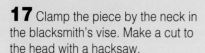

18

" Curl the tail on the face of the anvil, heating and hammering it gently. "

19

20

17 Clamp the piece by the neck in the blacksmith's vise. Make a cut to the head with a hacksaw.

18 Heat the cut area and separate the jaws of the mouth by hammering on a hardy.

19 Curl the tail on the face of the anvil by gently heating and hammering it on the rounded beak.

20 To refine the shape of the lizard, round the back by heating it and hammering it on the rounded beak of the anvil. The length of the hind leg has also been adjusted.

21 To make the nails and the base for attaching the knocker to the door, hammer the end of a rod until it is the same size in cross section as the hole in the heading tool.

22 After the nail is cut off the rest of the rod, heat and upset the part that will become the head. Repeat this to make the number of nails needed to attach the knocker to the door.

23 To hold the lizard in place, split a piece of previously tapered steel strapping along a cut made with an electric jigsaw. Drill a small hole at the end of the cut to make it easier to spread the sections apart.

24 Separate the two parts along the cut by heating and hammering on the rounded beak of the anvil.

25 The finished base plate and the fastening nails. Note that the spread ends of the base have been tapered. An electric drill was also used to make holes in the ends.

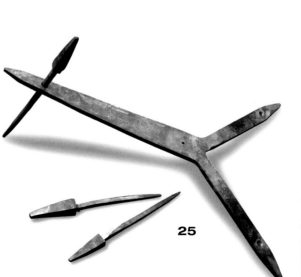

26 The knocker will strike this anvil. It is the end of a square rod, upset, slightly tapered, and then taken down to a smaller diameter with a grinder.

27 This part will be used to work the knocker. It consists of the tapered end of a square rod with two hacksaw cuts and a hole made with an electric drill. The material between the two cuts and the hole is removed to accommodate the back part of the lizard. Another hole has been made in one side to receive a pin with an upset at one end to serve as a pivot for the knocker.

28 Detail of the attachment of the two previous elements to the base plate on the door. The base plate has countersunk holes for inserting the ground-down areas; they are riveted after being heated and shaped.

29 Detail of how the knocker works.

The mounted door knocker. To protect it, a coat of varnish made of wax and colophony rosin has been applied. It evens out the colors and highlights the hand-forged texture.

Coat
rack

This project is a standing coat rack with a shape that is rather reminiscent of a plant. A model will be made out of sheet metal stock to check its functionality. Using an oxy-propane torch for the forging will demonstrate another way to heat the steel. Also, plasma cutting will be used to taper the ends of the sheet stock. Finally, the forged sheets will be welded together using coated rod.

1 and **2** Make scale drawings of the coat rack and of the pieces that compose it.

3 From the measurements in the drawings, create a full-scale model to make it easier to visualize the coat rack and how it works.

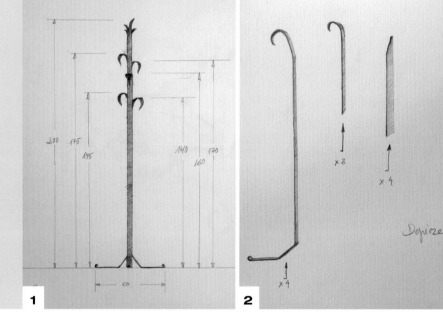

1

2

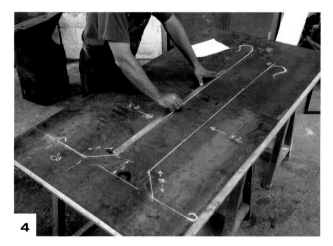

3

4 On a flat surface, such as a steel sheet, draw the outlines of each part of the coat rack at actual size. This will serve as a template for checking the work as you go along.

4

5 Forge the part that will become the foot of the coat rack. Heat the end with the flame of the oxy-propane torch.

6 Bend the heated end on the rounded beak of the anvil.

7 Finish bending the end by hammering it on the face of the anvil.

8 The finished foot of the coat rack after several heats.

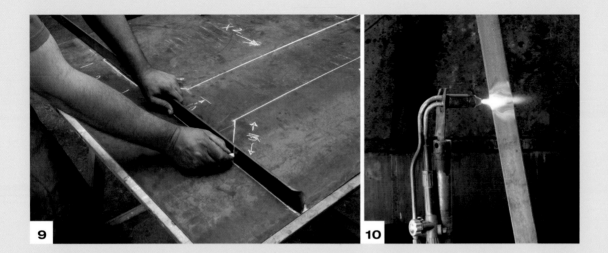

9 Place the forged strap onto the sheet-steel drawing, and use chalk to mark the place where it is to be bent.

10 Heat the marked area.

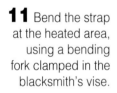

11 Bend the strap at the heated area, using a bending fork clamped in the blacksmith's vise.

12 The other bends are done in the same way: mark the pieces on the drawing, heat them, and bend them on the blacksmith's vise.

13 Throughout the construction process it's a good idea to keep checking the shape of the strap against the sheet-steel drawing.

14

15

16

17

18

14 Use the plasma cutter to remove part of the end of the straps to create a fairly tapered point.

15 Place the trimmed end over the face of the anvil and heat it with the torch flame. The strap can rest on a support stand while you heat it. Work carefully to avoid heating up the anvil.

16 Hammer the cut and heated end to produce a tapered, forged finish.

17 Heat it again, and curve the end of the strap on the rounded beak of the anvil.

18 Then, once the strap has cooled, give it a slight curve by levering it in a bending fork held in the jaws of the blacksmith's vise.

Step by step

113

19 Some of the forged pieces. They all match up to the sketched template.

19

20

21

22

23

20 The longest straps are welded to a piece of square tubing of the same width as the straps. The idea is to make two sets of strapping that fit together to form the coat rack.

21 The upper end of the coat rack, where the joining method is visible.

22 Join the two sets of straps together with a pin that passes through them. Weld the ends of the pin to the straps, and polish the weld to hide the welding beads.

23 To finish, go over all the welds with an angle grinder and a deburring disk. Note how the base is constructed to create stability and prevent wobbling.

Flat enamel paint was used to finish the coat rack.

Tricycle
stool

This project shows the steps for making a three-legged stool. The starting point is a stone with a shape that is similar to a bicycle seat; the stool is reminiscent of a tricycle. Three spirals of different sizes will be welded together using coated rod. We will also make a structure for holding the stone seat to the rod assembly.

1 and **2** Make a few preliminary sketches for planning purposes.

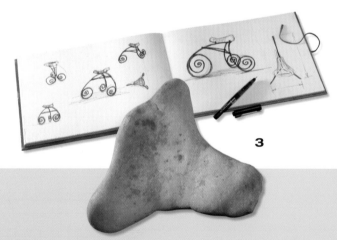

3 The saddle-shaped stone will be the main focus of the piece.

4 Draw a grid onto a copy of the original so that it can be enlarged to actual size without distorting the scale.

5 Draw a grid on a flat surface, using the same number of squares as in the sketch.

6 Transfer the shape by drawing freehand with chalk, keeping in mind the relationship between the grid and the original drawing.

7 During the transferring process, we noticed that the enlargement was too large for the stone, so we decided to reduce the size of the grid to remedy the problem.

8 Use a piece of rope on the spiral to determine how long the rod needs to be; the end is marked with chalk.

9 The steel rod is measured with the rope, and the distance is marked with chalk at the point where it will be cut.

10 Start the spiral by heating the end of the rod and hammering it on the face of the anvil.

"The leverage must be applied carefully to avoid forcing it too much and turning it into a bend."

11 Once the spiral is started, apply some leverage to it in a bending fork clamped between the jaws of the blacksmith's vise. The leverage must be applied carefully to avoid forcing it too much and turning it into a bend.

12 While the spiral is being bent, continue to check its shape against the drawing. Try to use the drawing as a guide, rather than copying it exactly.

13

14

13 Join the three completed spirals by welding a straight piece of rod of the same diameter with coated rod.

14 Curve a square rod by hammering it cold on a piece of U-shaped angle iron. This rod will serve to secure the spirals.

15 With the curved rod placed on the ends of the spirals, mark with chalk the places where it will be bent.

16 Use a torch to heat the spot where the bend will be made. Use a scrolling wrench as a lever while the metal is heated.

15

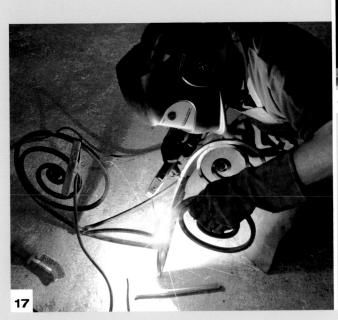

17

16

17 Place the now triangle-shaped rod on the ends of the three spirals, and weld in place. Wear safety gear to protect against the intense light and potential burns from the arc welding.

Step by step

119

18 Finally, construct the frame to hold the stone seat. Heat a piece of square rod, and bend it to follow the shape of the stone. With chalk, mark the point where the excess rod is to be cut.

19

19 The rods that form the support structure for the stone.

20 Close-up view of the joint in the support rods. Next, fill the hollows in the intersection between the rods with weld.

21 Set the support for the stone in place, and weld it to the part where the spirals are connected.

Stool shaped like a tricycle with the stone in place. The finish was made by applying a solution of water and salt, topped off with two coats of clear varnish.

Serpentine
weathervane

The following exercise takes you through the process of making a weathervane. The back part of the arrow is designed to reflect a cloth banner flapping in the wind. A special bending fork needs to be made, along with a template from thin wire that will serve as a guide in making the wavy shape. The tail itself will be forged from sheet steel about 1/16 inch (1 mm) thick. The pivot for the weathervane consists of an axis with a pointed end on which the center of the arrow pivots.

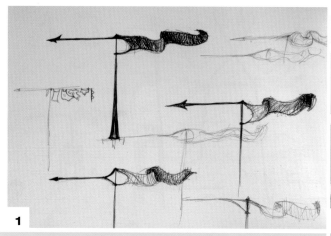

1

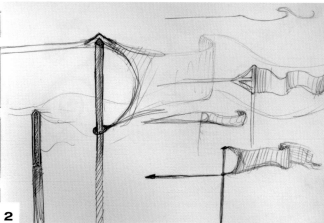

2

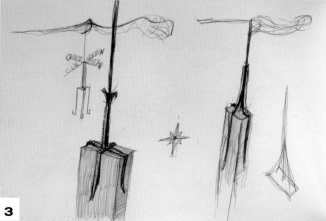

3

1 First make initial sketches of the weathervane. These show various possibilities being considered for the arrow's position. Each of them retains the idea of a cloth flapping in the wind.

2 Sketches are also drawn to work out support system and pivot for the body of the weathervane.

3 Preliminary sketches are also made to study how best to secure the axis of the weathervane to the wooden post where it will be mounted.

4 Bend a thin steel rod with a bending fork to model the approximate wavy shape for the end of the arrow.

5 Once the curve is created, cut off the rest of the steel rod.

6 Bend a round steel bar to create a bending fork that will make it possible to curve the entire width of the sheet stock. Because of its thickness, use a pipe to extend the lever and apply more force.

7 The finished bending fork. Its long legs make it possible to curve the entire width of the sheet stock for the back part of the weathervane's arrow.

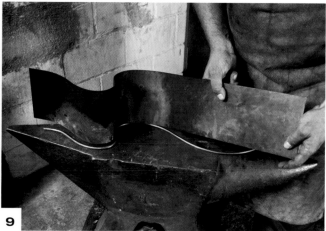

8 Heat the end of the sheet steel, and bend it by prying it in the bending fork you made in step 6.

9 While making the shape, periodically check it against the template.

10 Proceed very cautiously when heating a sheet that is only about 1/16 inch (1 mm) thick; it can easily reach high temperatures, and parts of the sheet may melt.

11 Continue working the sheet in the tool made for this project.

12 To finish, forge the end that will be connected to the front of the arrow by hammering it on the rounded beak of the anvil. The rest of the forged steel sheet is cooled in water so it can be handled.

13 To forge the arrow itself, make a tenon by hammering the end of a round rod on the edge of the anvil face.

14 Upset the tip a bit to thicken it so that the metal can be expanded to create the point of the arrow.

15 Reheat the upset end, and simultaneously hammer it out and taper it to create the point.

16 The shape of the arrow is worked cold with a triangular file. Clamp the bar securely in the blacksmith's vise so that you can bear down as you file.

Step by step

17 Heat the opposite end of the arrow and hammer it flat on the face of the anvil.

18 Shape the edges by turning the arrow as you hammer it out.

19 Place the heated rod over the round hole in the anvil face. Hit it with a pointed steel rod and a hammer to force the heated part into the hole in the anvil.

20 Next, make a slight twist by turning the flattened end with a scrolling wrench. This turn will later be used to connect the arrow to the forged steel sheet.

"This turn will be used to connect the arrow to the forged sheet."

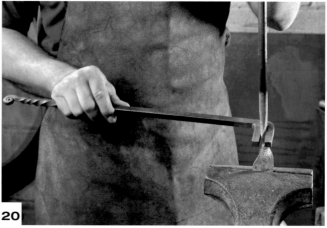

21 Connect the arrow to the wavy sheet steel by arc welding. Protect yourself against the radiation and ultraviolet rays that this type of welding produces.

22 A close-up view of the joint between the two parts of the weathervane.

23 and **24** Forge a ring that will stabilize the weathervane when it pivots on the main axis.

24

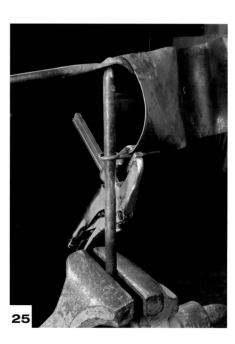

25 This view of the pivoting system for the weathervane clearly shows the twist and the indentation made in the hammered-out end of the arrow. Note the location of the ring for stabilizing the weathervane.

Step by step

127

26 Taper and round a square rod to forge the axis on which the weathervane will pivot.

27 After the end is forged, harden it to reduce the friction created by the rubbing of the weathervane against the axis.

28 Polish the rounded part that will be subjected to friction to make it easier for the device to pivot in the wind.

29 Forge the pieces for attaching the main axis to the wooden post, and bend them to the proper dimensions.

30 The attachment of the weathervane's axis to the wooden post.

The weathervane mounted on the wooden post.

Diàspora: a monumental
forging

We now will take a look at the forging process of part of the monumental sculpture *Diàspora* by the Spanish sculptor Ernest Altés. It is located in Copons (Barcelona, Spain) and is an homage to the mule drivers of the Anoia region who used to bustle along the roads of the Iberian peninsula. The sculpture was created at Irizar Forging in Lazkao. The techniques used for industrial-scale works are essentially the same as the techniques we have been exploring. They vary only in scale: the material is steel, but the hammer is a hydraulic press, the forge is a gas furnace, and the huge tongs are activated from a crane.

1 and **2** First, sketches and drawings that will guide the idea's development are created. One drawing leads to another, until it is time to put the idea into physical form.

3 Keep in mind that the thickness of the sheet steel and the various dimensions of the pieces will be reproduced based on the selected scale.

1

2

3

4

5

4 Inside the industrial forge, a cold-bent U-shaped sheet is put in the gas furnace to heat it. When pressure is applied to bend it, a couple of steel bars welded to the ends of the U maintain the shape.

5 With the help of huge mechanical tongs activated by a crane operator, the heated piece is taken out.

6 The piece is placed in the hydraulic press so that the area to be bent is close to the piston that will apply the pressure. Two pieces of steel placed on the floor just below the U-shaped piece can be seen; they will facilitate the bending by exerting pressure on the heated piece.

6

7

7 Detail of the piston of the hydraulic press that will apply the pressure on the heated piece. The rounded end that starts at the beads of weld was created specifically for making this project. A small triangle of steel welded to each side of the piston makes it possible to create the fold when the pressure is applied.

Step by step

8 The point at which pressure is applied to create the bend.

9 This photo clearly shows the steel rods welded to the end of the U shape to preserve the form during the bending process.

10 The angle of the bend is checked with a template made of sheet steel, which was fabricated to match the angle of the model.

11 The piece is placed back into the press to finish bending it to the correct angle.

12

13

14

12 Just as in traditional blacksmithing, the piece is heated as many times as necessary.

13 Cooling part of the piece with water during the forging process will keep it from distorting unnecessarily under pressure.

14 The bent piece is placed into the press once again, lined up with the piston, and pressure is applied until the desired shape has been achieved.

15 To finish the forging, localized heating is applied with an oxyacetylene torch. This repairs the small deformations that result from manipulating such a large piece.

"Just as in traditional blacksmithing, the piece is heated as many times as necessary."

15

Step by step

16

17

18

16 The final appearance of one part of the work after the forging process is complete.

17 Each part's three sections: two with forged bends, and one, the piece in the center, bent cold.

18 The dimensions are checked before connecting the three sections. Note the scale by comparing the artist, Ernest Altés, with the forged pieces.

The assembled sculpture *Diàspora* (2004), Copons, Spain. It consists of forged steel pieces and calcareous stone blocks from various locations in Catalonia. The sculpture is held together by a twisted steel cable.

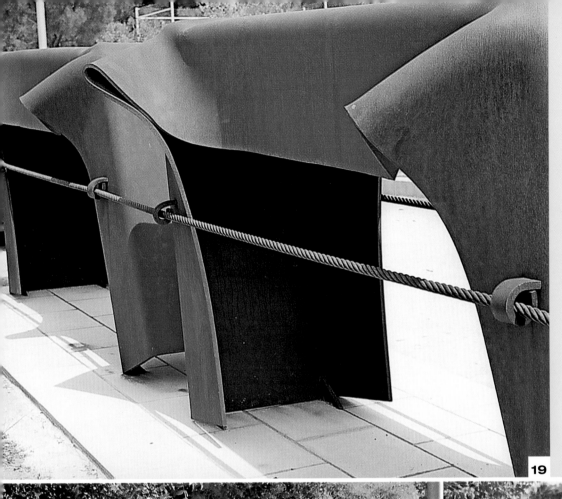

19 Close-up view of the forged bends. In applying the finish, the metal was sandblasted to remove forging marks and to even out the natural rusting. The rings were welded on, and the cable passes through them and secures the sculpture to the ground.

19

Gallery

Ernest Altés, *Ofrena*, 1995. Stone and forged steel. 3.9 × 6.3 × 6.7 in (10 × 16 × 17 cm).

Ares, *Xamán* ("Shaman"), 1996. Forged and welded steel. 16.9 × 5.5 × 4.7 in (43 × 14 × 12 cm).

Katherine Gili, *Bitter Joy*, 2005. Forged and welded steel. Height 59 in (150 cm).

Eduardo Chillida, *Peine del viento* ("Wind comb"), 1976. Paseo Eduardo Chillida, Donosti-San Sebastián, Spain.

Anonymous, Door knocker, 16th century, forged steel. # 31565, Museu Cau Ferrat, Sitges, Spain.

Anonymous, Grille, 20th century, forged steel. Marrakesh, Morocco.

Gallery

Ares, *Record per a un cop d'onada*, 2005. Oxy-cut and forged steel. 14.9 × 11.8 × 12.6 in (38 × 30 × 32 cm).

Ares, *Natura II*, 2003. Forged steel. 36.6 × 19.3 × 1.8 in (93 × 49 × 4.5 cm).

Anonymous, Door knocker, 19th century, steel. Old town Santiagomillas, León, Spain.

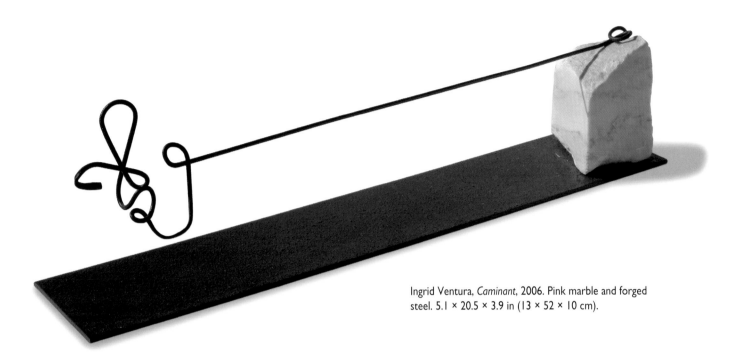

Ingrid Ventura, *Caminant*, 2006. Pink marble and forged steel. 5.1 × 20.5 × 3.9 in (13 × 52 × 10 cm).

Ernest Altés, *Ch'i II*, 1995. Basalt and forged stainless steel. 11 × 8.2 × 11 in (28 × 21 × 28 cm).

Antonio Sobrino and Mercedes Cano, *Bosque* ("Forest"), 2003. Forged steel. Various sizes. Museo del Hierro (Museum of Steel), Oropesa de Mar, Spain.

Gallery

139

Francisco Gazitua, *Huaso,* 2003. Forged steel.
11.8 × 11.8 in (30 × 30 cm). Private collection.

Anonymous, Shears.
Forged steel. Length
11.8 in (30 cm). Sonoran
desert, Mexico.

Anonymous, Steelyard balance, 20th
century, forged steel. Length 15.75 in
(40 cm). La Cerdaña, Catalonia, Spain.

Antoni Gaudí, *Barandilla de la Pedrera*, 1906–1910. Forged steel.
Casa Milà o La Pedrera, Barcelona, Spain.

Ares, *Petita remor*, 2000. Forged and oxy-cut
steel. 6.7 × 6.3 × 2.4 in (17 × 16 × 6 cm).

Gemma López, *Cabró ibèric*, 1999.
Forged steel. 11.8 × 11.8 × 15.75 in
(30 × 30 × 40 cm).

B- Bead. A deposit of metal that unites two pieces of metal in the welding process.

C- Cast iron. Alloy composed of iron, carbon (in a proportion of over 1.7%), and silicon, used for producing pieces in molds.

Cold chisel. Tool for cutting cold or hot steel with the blow of a hammer.

Colophony rosin. A residual product from the distillation of essence of turpentine.

Contraction. A reduction in volume of a metal after expanding, e.g., as it cools down.

Corrosion. Chemical reaction that involves the slow destruction of the metal through the action of external agents.

D- Double boiler. A process in which moderate, continuous heat is applied with a burner; one container, containing the material to be heated, is located inside another water-filled container, and is heated by conduction from the surrounding water.

Ductility. A property of metals that makes it possible to stretch them without breaking.

E- Electric intensity. The quantity of electrons that pass through a section of conductor in a unit of time.

Expansion. An increase in the volume of metal through the effect of heat.

F- Fusibility. The ease with which a metal turns to liquid through heat absorption.

Fusion. The change of a solid body into a liquid due to heat.

Fusion bath. In welding, the space where the heat causes the added metal and the metal base to melt, forming a volume of material that produces the welding bead when it cools.

G- Gum lacquer. A natural resin that is sold in the form of flakes.

Glossary

H- Hardness. The resistance of metals to wear by friction.

Heating. The action of heating metal in the fire of the forge or by other means, such as a torch.

I- Iron. A strong, hard metal with a carbon content of less than 0.05%.

M- Malleability. The capacity of heated metal to change shape through hammering.

Manometer. A valve for reducing the pressure of a compressed gas in a tank or a compressor to the optimum working pressure.

Mild steel. A steel alloy containing little carbon; ideal for blacksmithing work, but not suited to heat treating such as hardening.

O- Oxidation. The combination of metal with oxygen from the air, forming a layer of oxide or rust.

P- Patina. A fine layer of oxide deposited on the surface of metals.

Plumbago. Powdered graphite used for making artificial patinas.

S- Scroll. A spiral-shaped figure.

Slag. In welding, a material produced by the decomposition of the coating on coated rod. Its purpose is to protect the fusion bath and the weld from oxidation.

Spot welding. The action of joining two pieces of metal through welds at various points.

Stainless steel. Alloy of steel, chrome, and nickel, among other metals, that is particularly resistant to corrosion.

Steel. An alloy made up basically of iron and carbon. The carbon percentage is between 0.05 and 1.7%.

Tenon. A protrusion in the shape of a piece of steel after drawing it out while hammering it at heat.

V- Vacuum voltage. The voltage in the ground clamp and the electrode clamp when not welding.

Acknowledgments

To Parramón Publishing for the creation of this and other resources dedicated to the arts and crafts.

To the University of Barcelona for allowing the use of its space to develop the projects.

To editors María Fernanda Canal and Tomàs Ubach for their trust and patience with the author.

To Joan Soto, photographer, for his outstanding professionalism, his suggestions, and his good humor.

To the sculptor Ernest Altés for his great disposition and attention, as well as for providing the photos of the making of his creation *Diàspora*.

To Rafael Cuartiellas, workshop foreman and artist, for his subtle teachings.

To the sculptors:
Ernest Altés, Gemma López,
Ingrid Ventura, Antonio Sobrino, Mercedes Cano, Katherine Gili, and Fernando Gazitua, for providing images of their artistic work for the gallery section.

To Valerià Cortés for her inestimable help in the step-by-step project of the grille with scrolls.

To Jordi Torras and Rubén Campo, shop masters, for their daily assistance in transporting items and locating tools and materials for the projects.

To D. José Ares, from Valdespino de Somoza, León, Spain, for allowing me to photograph his blacksmith shop.

To my parents, Clavelina and José, for taking care of my girls so many times while I worked on this book.

I especially would like to thank Marta and our daughters la and Ona (in the photo) for their help and understanding while this book was being created. It would not have been possible without their laughter.

144